AF127649

FREE RADICALS IN MEDICINE

FREE RADICALS IN MEDICINE

RADU OLINESCU
AND
TERRANCE SMITH

Nova Science Publishers, Inc.
Huntington, New York

Senior Editors: Susan Boriotti and Donna Dennis
Coordinating Editor: Tatiana Shohov
Office Manager: Annette Hellinger
Graphics: Wanda Serrano
Book Production: Matthew Kozlowski, Jonathan Rose and Jennifer Vogt
Circulation: Cathy DeGregory, Ave Maria Gonzalez, Ron Hedges and Andre Tillman

Library of Congress Cataloging-in-Publication Data
Available upon request.

ISBN 1-56072-869-8

Copyright © 2002 by Nova Science Publishers, Inc.
        227 Main Street, Suite 100
        Huntington, New York 11743
        Tele. 631-424-NOVA(6682)   Fax 631-425-5933
        e-mail: Novascience@earthlink.net
        Web Site: http://www.novapublishers.com

All rights reserved. No part of this book may be reproduced, stored in a retrieval system or transmitted in any form or by any means: electronic, electrostatic, magnetic, tape, mechanical photocopying, recording or otherwise without permission from the publishers.

The authors and publisher have taken care in preparation of this book, but make no expressed or implied warranty of any kind and assume no responsibility for any errors or omissions. No liability is assumed for incidental or consequential damages in connection with or arising out of information contained in this book.

This publication is designed to provide accurate and authoritative information with regard to the subject matter covered herein. It is sold with the clear understanding that the publisher is not engaged in rendering legal or any other professional services. If legal or any other expert assistance is required, the services of a competent person should be sought. FROM A DECLARATION OF PARTICIPANTS JOINTLY ADOPTED BY A COMMITTEE OF THE AMERICAN BAR ASSOCIATION AND A COMMITTEE OF PUBLISHERS.

Printed in the United States of America

Contents

Forward		vii
Chapter 1	**Introduction to Free Radicals**	1
	1.1 Definition of Free Radicals	1
	1.2 Formation of Free Radicals	3
	1.3 Dynamics of Free Radical Formation	6
	1.4 Methods to Measure Free Radicals	8
Chapter 2	**Types of Non-Oxygen Free Radicals**	11
	2.1 Flavine Semiquinones	11
	2.2 Quinones and Semiquinones	16
	2.3 Aromatic Compounds	17
	2.4 Nucleic Acids	19
	2.5 Thylil Radical	20
	2.6 Nitric Oxide	21
Chapter 3	**Oxygen Free Radicals**	23
	3.1 The Oxygen Paradox	23
	3.2 Singlet Oxygen	25
	3.3 Superoxide	26
	3.4 Hydrogen Peroxide	28
	3.5 Hydroxyl Radical	29
	3.6 Peroxides	31
Chapter 4	**Natural Processes as Sources of Free Radicals**	35
	4.1 Mitochondrial Respiration	35
	4.2 Peroxisomes	38
	4.3 Phagocytosis	40
	4.4 Prostaglandins and Leukotrienes	44
	4.5 Peroxidation of Unsaturated Lipids	47
	4.6 Are Oxygen Free Radicals Chemical Messengers?	51
	4.7 Apoptosis	52
Chapter 5	**Harmful Effects of Free Radicals**	53
	5.1 Oxidative Stress	53
	5.2 Cell Membrane Damage	58
	5.3 Amplifying Factors	61

Chapter 6	**Radiation and Free Radicals**	71
	6.1 UV Radiation	71
	6.2 Ionizing Radiation	80
	6.3 Eye Diseases	84
Chapter 7	**Antioxidants**	87
	7.1 Introduction to Antioxidants	87
	7.2 Natural Antioxidant Enzymes	90
	7.4 Mineral Antioxidants	113
	7.5 Plant Antioxidants	116
	7.6 Synthetic Antioxidants	118
Chapter 8	**Physiologic Effects of Oxygen Stress**	119
	8.1 Introduction	119
	8.2 Pregnancy and the Neonate	119
	8.3 Physical and Emotional Stress	120
	8.4 Aging	121
	8.5 Hyperbaria	126
	8.6 Chemical Pollution	127
Chapter 9	**Free Radicals and Pathology**	129
	9.1 Inflammation and Rheumatic Conditions	129
	9.2 Cardiovascular Disease	135
	9.3 Cancer	142
	9.4. Respiratory Failure	144
	9.5 Liver Failure	149
	9.6 Physiological Brain Diseases	151
	9.7 Others	154
Chapter 10	**Therapeutic Drugs and Nutrition**	157
	10.1 Introduction	157
	10.2 Prooxidant Drugs	158
	10.3 Antioxidant Drugs	160
	10.4 Lipids	162
	10.5 Free Radicals, Alterative Medicine, and Diet	165
	10.6 Antioxidant Therapy	170
Cited References		173
Index		185

Forward

The role of free radicals in normal and disease processes has become a major area of interest in the scientific community. In about 30 years this area of study has developed from a state of rejection by reputable scientist to general acceptance to intense interest and study. However, textbooks have not yet begun to cover the field of free radical biological effects in detail. The student of free radicals in biology must glean his background information from widely dispersed sources: old journal articles, proceedings of symposia, etc.

Free Radicals in Medicine seeks to alter this situation. This book provides an indepth overview of the nature, biological and environmental formation, deleterious effects on cells, and theraputic uses of free radicals. The book is intended for use in the university classroom as a supplementary text in the study of biochemistry.

Chapter 1

INTRODUCTION TO FREE RADICALS

1.1 DEFINITION OF FREE RADICALS

The cell is the basic unit of life, and its complexity still is capable of amazing us. The understanding of all of the physicochemical processes that take place within the cell is a task that remains incomplete. However, it can be stated that within the cell these processes result in the maintenance of the structure and function of the cell.

According to the second law of thermodynamics, all systems tend toward equilibrium. In the case of the cell this is a dynamic equilibrium resulting from the coupling of exergonic and endergonic reactions. These reactions serve to constantly renew the structural components of the cell and counteract the destructive forces that impact the cell from without and within. In addition, these reactions are central to the production of energy to power the cell and to the production of compounds that function to maintain interaction between the individual cell and the rest of the body.

While the biochemical activity within the cell is highly differentiated, stability of the reaction pathways and their products is important to the normal function of the cell. At the same time, the cell requires flexibility in response in order to adapt to newly encountered harmful agents such as microbes and toxins.

This tension between stability and flexibility is made possible in the cell through enzymatic reactions. These reactions allow fast cycle times and most are susceptible to modification or regulation by other compounds, often a reaction product. Once the importance of enzymes and enzymatic reactions to cellular function was understood it became possible to recognize the importance of free radicals in metabolic processes.

It is now recognized that free radicals play an important role in numerous enzymatic and nonenzymatic reactions within the cell. For example, free radicals are formed as necessary intermediates in some oxido-reductive reactions. In these reactions rapid structural modifications take place that often have significant effects on the compound's biological activity. Free radicals are also important in the destruction of invading microbes, intracellular communication, secondary messengers, etc.

Dr. William Pryor, a leading expert in free radical research, has pointed out the controversial nature of free radical involvement in cellular activity [140]. At the same time it must be noted that the concept of free radicals and free radical reactions has been stated as early as the eighteenth century. In 1896 the chemist Oswald stated that free radicals cannot exist in nature because they are so reactive they have a very short half-life. Then, in 1900, Gombeth for the first time isolated a relatively stable free radical, triphenyl-methyl: $(C_6H_5)_3C^+$. Between 1926 and 1929, Paneth isolated methyl and ethyl free radicals following thermal decomposition of tetramethyl lead.

$$Pb(CH_3)_4 \longrightarrow Pb + 4CH_3^{\bullet} \text{ (methyl radicals)} \tag{1}$$

For a long time these organic free radicals were nothing more than a curiosity. Only in 1939, when Michaelis demonstrated enzymatic oxido-reductive reactions, were some free radicals recognized as being an enzymatic intermediate. Today it is recognized that intermediate free radicals in oxido-reductive reactions have a short, but variable half-life (from a thousandth of a second to several minutes). These discussions on free radicals that started a century ago are not yet concluded due to the complexity of biological systems, the high reactivity of free radicals, and to the variety of experimental conditions.

Free radicals are defined as molecules that contain an unpaired electron. Organic molecules normally possess an even number of electrons with each orbital being occupied by two electrons having an opposite magnetic moment and spin. Consequently, free radicals are characterized by the following.

a. Contain one or more unpaired electrons
b. Have variable electronic charge. They may be neutral, electropositive or electronegative. To denote this in equations, the free radical state is indicated with a dot next to the element carrying the radical.
c. Possess such a great chemical reactivity that rate constants can reach the limit for free diffusion. ($10^9 M^{-1} s^{-1}$). Generally, the simpler the structure, the shorter the life of free radicals. Complex molecules, like triphenlymethyl, have more stability and a longer life span.
d. Appear in a variety of conditions and can participate in interchange reactions.

Examples of the last point are shown here. The drug chlorpromazine may react with an oxygen radical (hydroxyl radical in this example) to produce a chlorpromazine cation radical and a hydroxy group (non-radical).

$$\text{Chlorpromazine (CPZ)} + OH^{\bullet} \longrightarrow CPZ^{\bullet} + OH^{-} \tag{2}$$

Under some conditions, oxygen may capture an electron producing the superoxide anion, the most basic oxygen free radical (3), or carbon tetrachloride may capture an electron, producing a neutral free radical, trichloromethyl ($CCl_3^{\bullet}$) (4).

$$O_2 + e^{-} \longrightarrow O_2^{\bullet} \text{ (superoxide)} \tag{3}$$
$$CCl_4 + e^{-} \longrightarrow CCl_3^{\bullet} + Cl^{-} \tag{4}$$

These free radicals have great chemical reactivity, possessing rate constants between 10^4 and $10^9 M^{-1} s^{-1}$. This means that, in effect, these reactions proceed nearly instantaneously.

It is interesting to note that much of the recent research into free radical reactions is a consequence of the cold war. Following the invention of nuclear weapons in the second world war, there was a great deal of interest in radiation disease. The discovery that free radicals are involved in the first steps of the consequences of irradiation of living organisms attracted much attention to the field of free radical research. Likewise, the interest in free radicals by physicians and medical researchers is based on their role in physiology and pathology.

1.2 FORMATION OF FREE RADICALS

Free radicals are formed in chemical reactions that take place everywhere; in the atmosphere or on the earth, in inert material and in living organisms. Among this huge group of reactions there are some that are of particular interest to us in examining biological systems.

A. Homolysis of compounds by heat. This occurs with fatty acids, some of which can generate free radicals at moderate temperatures, triggering further free radical reactions (reaction 5).

$$AB \longrightarrow A^{\bullet} + B^{\bullet} \tag{5}$$

These initiators (lipid hydroperoxides) have a weak, but reactive, oxygen to oxygen bond. These type of reactions occur in rancidification reactions in fats and oils and is part of the polymerization process of drying oils.

B. Coupled lysis. Hydroperoxides decompose rapidly in the presence of some compounds, especially alcohols, acids, and other peroxides. Coupled lysis of enzymatically formed hydroperoxides is part of the synthesis reaction for prostaglandins.

C. Electron transfer takes place in oxido-reductive reactions and is the most common mechanism for the formation of free radicals, especially those of oxygen. These reactions take place through metal ion-induced hydrogen peroxide decomposition. Iron ion is the best known inducer and the Haber-Weiss cycle (reactions 6 and 7) has been demonstrated under numerous experimental conditions attempting to show the *in vivo* formation of lipid peroxides.

$$H_2O_2 + Fe^{2+} \longrightarrow OH^{\bullet} + OH^{-} + Fe^{3+} \tag{6}$$

$$H_2O_2 + Fe^{3+} \longrightarrow HO_2^{\bullet} + H^{+} + Fe^{2+} \tag{7}$$

The Haber-Weiss cycle may also function with copper ions or with organic complexes of iron (EDTA-Fe). The decomposition of peroxides under the catalytic action of hemoglobin is also part of this type of reaction.

D. Photolysis and radiolysis. Important amounts of free radicals are produced following exposure to ultraviolet light or ionizing radiation. The yield of free radicals under these conditions depends on factors such as wavelength of the light, time of exposure, etc. While UV light acts only on those compounds that favor light absorption, ionizing radiation is capable of destroying or damaging any organic molecule found in an organism.

While the action of UV is limited to the period of exposure, the action of ionizing radiation continues after exposure ends due to the radiolysis of water. Since water constitutes over 70% of living organisms, this continuing reaction can be of great importance. Radiolysis of water is summarized in reaction (8).

$$H_2O \longrightarrow (H_2O^+, e^-) \longrightarrow H_3O^+, H^{\bullet}, OH^{\bullet}, e^-aq, H_2O_2, H_2 \tag{8}$$

There is a high yield of free radicals with the radiolysis of water, especially $OH^{\bullet}$ (hydroxyl) and $H^{\bullet}$ (proton), and reactive species like e^-aq (hydrated electron) and H_2O_2 (hydrogen peroxide). Most of these are very reactive and capable of damaging organic molecules within the body.

E. Enzymatic reactions are a continuous source of free radical intermediates. These free radicals either decompose or bind with other, stable, compounds. The most frequently formed organic free radicals are the semiquinones, summarized in reaction (9).

$$\text{oxidant + reductant} \rightarrow \text{2 free radicals (semiquinones)} \rightarrow \text{final products} \tag{9a}$$

$$\text{benzoquinone + NADH} \xrightarrow{\text{diaphorase}} \text{semiquinones} \longrightarrow \text{hydro-quinones + NAD} \tag{9b}$$

The maximal yield of free radicals are produced when the reactants are in equimolar concentrations. The rate of free radical formation (V_f) is given in equation (10).

$$V_f = KV = 2Ko[AH^{\bullet}]^2 \tag{10}$$

Thus, the rate of free radical formation is proportional to the square of their concentration and to the square root of the enzyme concentration. By ESR and spectrophotometric studies it has been demonstrated that the life of free radical-enzyme complexes is approximately 10^{-10} seconds. In a redox enzymatic reaction (reaction 11), the number of electrons transferred varies as a function of the reactant structure with one electron for acceptors and two electrons for donors. This is not absolute but varies with enzyme type.

Electron acceptor: $AH_2 \xrightarrow{\text{enzyme}} AH^{\bullet} \longrightarrow A$ (oxidation reaction) (11a)

Electron donor: $A \xrightarrow{\text{enzyme}} AH^{\bullet} \longrightarrow AH_2$ (reduction reaction) (11b)

Peroxide is an acceptor, while flavine enzymes act as electron donors. In table 1.1, some enzymatic redox reactions are presented. This table illustrates that compounds with different chemical structure may be metabolized by several mechanisms. Second, there is the possibility of adaptation of biological systems. For example, reactions by peroxidase, ascorbate oxidase or xanthine oxidase may be involved in the modification of a wide range of compounds with related aromatic structures.

F. **Metabolism** occurs of organic compounds, especially those that easily penetrate the cell such as carbon tetrachloride (CCl_4) or chloroform ($CHCl_3$). Exposure to these compounds may be due to pollution or industrial accidents. These solvents react with proteins or amines (RNH_2) in the cell.

$$CCl_4 + RNH_2 \longrightarrow (RNH_2^+, Cl^-) \longrightarrow RNH_2^+ + Cl^- + C^{\bullet}Cl_3 \qquad (12)$$

The free radical, $C^{\bullet}Cl_3$, produces hepatic damage including cirrhosis. Secondly, again through metabolism, many carcinogenic hydrocarbons produce intermediate free radicals. This reaction of polyaromatic hydrocarbons such as benzanthracene and methylcholantrene will be discussed in chapter 2. Cytostatic drugs, such as adriamycin, can also produce free radicals.

G. **Free radical in nature.** Only in the past 80 years has the presence of free radicals in nature been demonstrated, especially in the atmosphere. Nitrogen and oxygen easily form specific free radicals that generate other free radicals through reactions with other compounds. For example, NO and NO_2 may be included among the stable free radicals present in the atmosphere, although in very small amounts (0.2 ppm NO_2 is present in smog). Relatively recently, stable aromatic free radicals were found in some volcanic rocks. They were probably formed under special conditions existing during volcanic eruptions.

Table 1.1. The principle redox reactions that form free radicals

Electron transfer	Molecule oxidized	Enzymatic reaction
1e (K=2)	Ascorbic acid, hydroquinone, indoleacetic acid, NADH	Peroxidase + H_2O_2
1e (K=2)	Ascorbic acid, hydroquinone	Ascorbate oxidase + O_2 Lactase + O_2
Mixed (K=0-2)	Sulfide, iodides	Peroxidase + H_2O_2
2e (K=0)	Catechol, NADH, D-amino acid	Tyrosinase + O_2 Dehdrogenases + acceptor D-amino acid oxidase + O_2
1e (K=2)	p-benzoquinone, menadione (vitamin K_3), oxygen	NADH-cytochrome b5 reductase NADPH cytochrome c reductase Xanthene dehydrogenase Xanthene oxidase + xanthene Aldehyde oxidase + aldehyde
2e (K=0)	Menadine, oxygen	Diaphorase + NADH Glucose oxidase + glucose

1.3 DYNAMICS OF FREE RADICAL FORMATION

We have given a brief presentation of free radical formation as it occurs in nature. Free radical formation is a complex, continuous process that, if no inhibitors are present, can continue as an indefinite chain reaction. This chain reaction consists of distinct phases that each last a variable period of time, depending on local conditions. So, starting with an organic compound, RH, we can distinguish the following phases.

A. Initiation. The compound loses an electron to form a free radical.

$$RH \longrightarrow R^{\bullet} \tag{13}$$

Once a free radical ($R^{\bullet}$) is formed, it can randomly react with another free radical (reaction 14), but will more likely react with oxygen (reaction 15).

$$R^{\bullet} + R^{\bullet} \longrightarrow RR \longrightarrow \text{stable or unstable products} \tag{14}$$

$$R^{\bullet} + O_2 \longrightarrow RO_2^{\bullet} \tag{15a}$$
$$RO_2^{\bullet} + RH \longrightarrow ROOH \text{ (peroxide)} + R^{\bullet} \tag{15b}$$

The formation of oxygen free radicals is one of the most important consequences of free radical generation. Oxygen free radicals are the main source of tissue damage, and as will be discussed in chapter 6, are the actual damaging agents with which the cell must cope.

B. Propagation. The propagation of free radicals is a factor that is most important in living systems. Outside the cell or body, a newly formed free radical will quickly decompose or react with oxygen. But, in living organisms, once a free radical is formed, it can react not only with oxygen but with proteins, lipids, or carbohydrates. During propagation the possible reactions are nearly endless.

Transfer of Atoms (mostly hydrogen or chloride)

$$R^{\bullet} + ACl \longrightarrow RCl + A^{\bullet} \tag{16}$$

Such reactions (SH_2 type) take place during the metabolism of polluting chlorinated hydrocarbons. The same type of reaction takes place with the sulfur atom following exposure to ionizing radiation. In such cases, enzymes possessing a free sulfhydryl group (SH) will be affected, resulting in a thyil free radical (RS˙), which will inactivate the enzyme.

$$Enz\text{-}SH + R^{\bullet} \longrightarrow Enz\text{-}S^{\bullet} + RH \tag{17}$$

Addition Reactions

Free radicals have a great affinity toward any unsaturated bond such as exists in unsaturated fatty acids. The addition of a free radical to a double bond provides a rapid propagation of new free radicals. This has been demonstrated in living organisms and been found to involve lipids or nucleic acids (nitrogen bases).

$$R^{\bullet} + LH \text{ (lipid)} \longrightarrow RH + L^{\bullet} \tag{18}$$

The reactions of free radicals with unsaturated fatty acids and with purine or pyrimidine bases of nucleic acids takes place at extremely rapid rates that are close to the diffusion rate ($10^{10} M^{-1} s^{-1}$).

This mechanism of propagation is also at work in manufacturing processes such as the aging of polymers, rubber, preserved foods (rancidification), etc.

C. Termination. As will be presented later in this book, free radical generation is a random event. Most free radicals then quickly decompose by reacting with surrounding compounds. Living organisms are equipped with varied and efficient protective mechanisms, called antioxidants. These reactions with antioxidants end the propagation reactions.

The auto-oxidation of lipids leading to the formation of peroxides is an excellent example. Polyunsaturated fatty acids, such as arachidonic acid, are main targets of free radicals because of the large number of double bonds. The process begins as demonstrated in reactions 13 and 18. The newly formed free radicals can react as demonstrated in reaction 14 and decompose to stable products. However, most free radicals will react with oxygen (reaction 15) forming peroxides. In the absence of antioxidants, free radicals might react with each other, terminating the chain reaction.

$$R^\bullet + ROO^\bullet \longrightarrow ROOR \text{ (stable product)} \tag{19a}$$

$$2\, ROO^\bullet \longrightarrow O_2 + ROOR \text{ (stable product)} \tag{19b}$$

Under standard conditions, the concentration of oxygen in air is approximately 10^{-3} M. Under these conditions, peroxidation will terminate as demonstrated in reaction 19. Under ischemic conditions the decreased oxygen concentration will allow the free radicals to accumulate and they will have opportunity to react with cellular constituents leading to major damage to membrane structure and cell function.

Finally, once started, and in the absence of antioxidants, the formation of free radicals will lead to complex peroxidative reactions that are difficult to predict. The concurrent presence of oxygen, lipids, and metallic ions presents favorable conditions for the peroxidation of cellular membranes. This is a major event following exposure to ionizing radiation or toxic compounds. With the decomposition of peroxides, new reactive compounds appear, such as aldehydes, ketones, organic acids, alcohols, epoxides, etc. that continue the oxidative destructive action of free radicals. Consequently, once a free radical reaction is started, it will continue and spread destruction. To protect against this total destruction of the cell or organism, antioxidants are present in tissues.

1.4 METHODS TO MEASURE FREE RADICALS

One reason knowledge about free radical has been slow in developing is the lack of adequate methods for quantification of their concentration and effects. However, in recent years, both direct and indirect methods have been developed.

A. **Spectrophotometry** is the oldest and best known method for analysis of free radicals. It was successfully used by B. Chance and J. Massey in what are now considered classical studies concerning the kinetics and mechanisms of some antioxidant enzymes, such as catalase, peroxidase, and flavine enzymes. These studies were performed using specialized spectrophotometers having high resolving power and able to record high-speed spectral changes. Such instruments are now commonly available. Absorption spectroscopy requires the use of compounds with high molar absorption coefficients and that react to form stable end products. In spite of these limitations, this method is still used widely, especially when studying the coupling of specific inhibitors.

B. Chemiluminescence involves the detection of nonvisible wavelengths of light emitted from some reactions in which a great deal of energy is released. Such exergonic reactions are common among reactive oxygen species generating systems. The emissions occur when released energy is capable of producing electron excitation, with the luminescence occurring as the electrons relax to ground state. Chemiluminescence is measured with scintillation counters or special instruments with sensitive photomultipliers. This is not a direct measurement of free radicals, but rather measures the consequences of free radical formation on the surrounding system.

C. Electron Spin Resonance (ESR) is a direct method for measurement of free radicals that is possible because free radicals possess a paramagnetic nature. When an external magnetic field is applied, paramagnetic compounds produce specific signals that can be interpreted quantitatively and qualitatively. The g factor provides quantitative information concerning the magnetic moment of electrons. Free radicals possess a characteristic hyperfine structure resulting from the interaction of nuclei and the electron orbitals. ESR is the most widespread method used to study free radicals, but is only useful for analysis of simple systems. Its use in biological systems is limited by the amount of water present in tissues (about 70%). Water possesses an magnetic dipole and therefore, quenches ESR signals. Consequently, when tissues are to be analyzed by ESR, they must be lyophilized and the measurements are made at the temperature of liquid nitrogen.

In spite of these limitations, ESR has had many successes in elucidating reactions in biological systems. Among these results are the evaluation of supposed free radicals present in cancer tissue. Following initial indications that free radicals were present at high concentrations in tumors, it was realized that the "free radical like ESR signals" were being produced by the higher coupling of electron orbitals of metallic ions found in the samples. Later, it was observed that such ESR signals occur in all tissues or cells that were isolated after metabolic activity had been stopped. It was then thought that the ESR signals that appear in cancer tissue were produced by mono-dehydroascorbate free radicals that survived the tissue isolation process. Finally it was found that under certain conditions, hemoglobin or iron atoms react with nitric oxide (NO) present in the tissues. Nitric oxide exhibits several properties of free radicals and when it reacts with compounds containing a heme group (hemoglobin, cytochrome, peroxidase) reactive compounds result.

ESR methodology has been enriched by the technique of spin trapping. This technique involves the reaction of free radicals with compounds called spin markers, producing adducts or specific compounds. Spin markers possess double bonds that are targets for binding free radicals.

$$X=Y \text{ (spin marker)} + R^{\bullet} \longrightarrow R\text{-}X\text{-}Y^{\bullet} \text{ (adduct)} \tag{20a}$$

$$R^{\bullet} + A\text{-}HC=CH\text{-}A' \longrightarrow (R\text{-}CH\text{-}CH)^{\bullet} \text{ (rapid reaction)} \tag{20b}$$

These reactions favor the accumulation of adducts possessing free radical properties, allowing low levels of free radicals to be measured.

Spin markers coupled with ESR were successfully applied to studies on liver microsomes. In these studies, the metabolism of xenobiotic compounds, aromatic hydrocarbon free radicals, lipid peroxides, and oxygen free radicals by the microsomal electron transport system has been examined. This has helped improve our understanding of the pathological consequences of free radical formation in the body.

Chapter 2

TYPES OF NON-OXYGEN FREE RADICALS

2.1 FLAVINE SEMIQUINONES

Semiquinones (SQ) were the first free radicals discovered to be involved in biological processes. Their properties have been known for over 30 years [142, 192]. To understand the role of SQs, we must review some basic theory.

Oxidation-reduction (redox) reactions are the main sources of energy for organisms. In such reactions, the oxidant (A) accepts electrons while the reductant (D) donates electrons. Therefore, in enzyme-catalyzed reactions the substrates in the reduced state (DH_2) are oxidized through electron transfer.

2 equivalents of redox: $DH_2 + 1/2\ O_2$ —enzyme→ $D + H_2O$ (1)

4 equivalents of redox: $2\ DH_2 + O_2$ —enzyme→ $2\ D + 2\ H_2O$ (2)

All enzymes involved in oxidation and reduction reactions are designated as oxidoreductases or oxidases. Due to the large redox potential between oxidative and reductant compounds, few enzymes are able to directly reduce oxygen to water. Therefore, living systems contain electron transport systems such as those illustrated in figure 2.1. The simple interactions between electron donors (D) and acceptors (A) (figure 2.1 A and B) are built into more complex systems like the electron transport chain (figure 2.1 C)

In addition to redox reactions, there are other reactions in which only one electron is transferred.

$AH + O_2 \longrightarrow A + HO_2^{\bullet}$ (an oxygen free radical) (3)

These univalent reductions are not widespread because they produce oxygen free radicals.

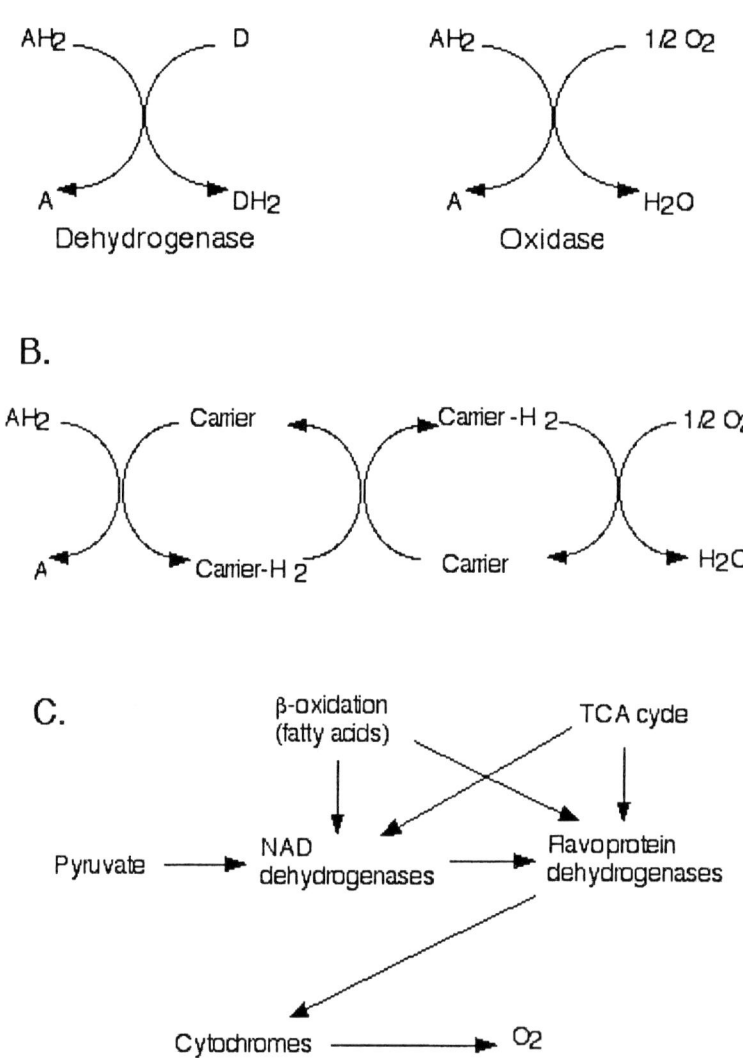

Figure 2.1. Examples of enzymatic redox reactions. A, Dehydrogenase and oxidase; B, A chain of redox reactions illustrating the principle of electron transport in the mitochondria; C, Steps that occur in the mitochondrial electron transport chain.

Aerobic dehydrogenases are flavoprotein enzymes containing the coenzyme flavine mononucleotide (FMN) or flavin adenine dinucleotide (FAD). The chemical structure of FAD, FMN, and the formation of flavin semiquinone free radicals are shown in figure 2.2. Flavin-enzymes play an important role in mitochondrial electron transport. Flavoprotein dehydrogenases have a crucial place between NAD dehydrogenases and cytochromes. Flavine-enzyme action involves several complex mechanism in which free radical semiquinones are formed. Flavine dehydrogenases are very important as they catalyze dehydrogenation of several substrates through reaction (4).

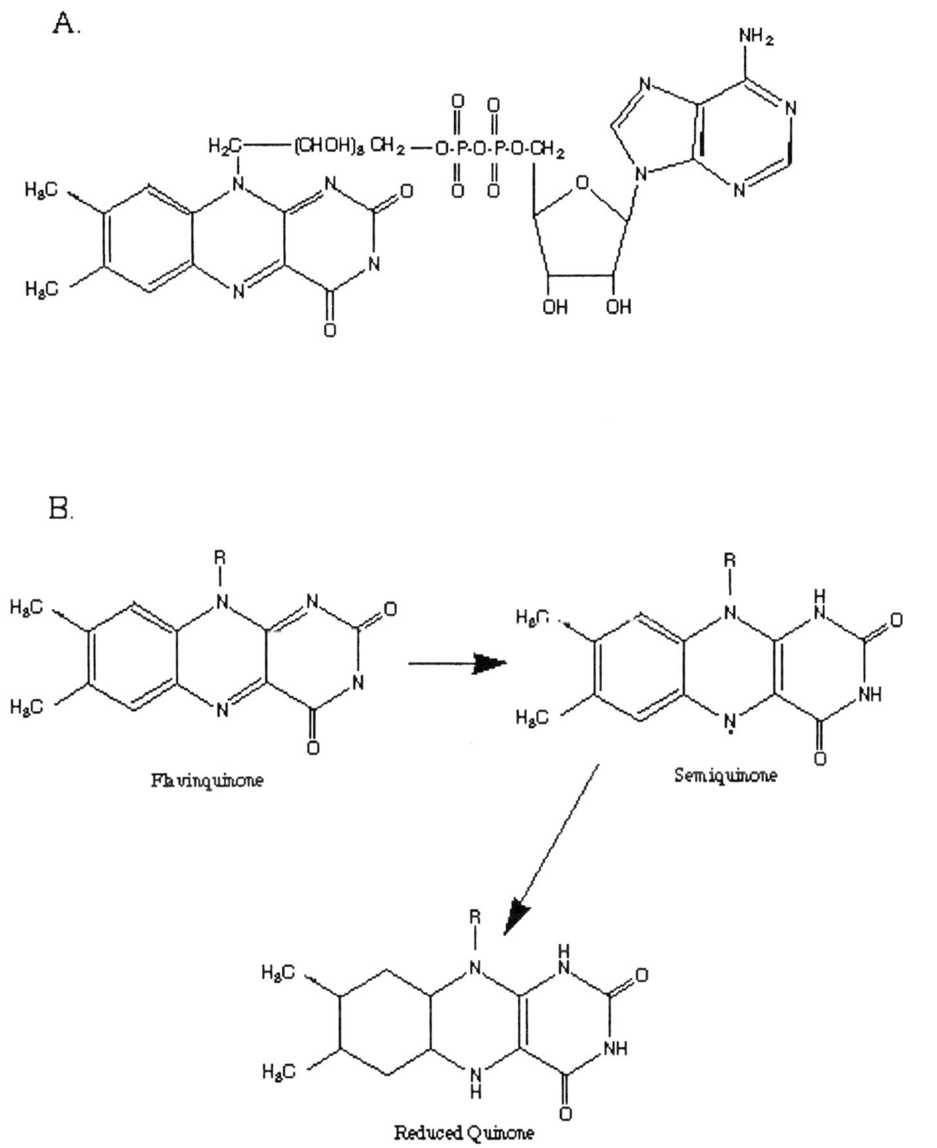

Figure 2.2. A, Flavine adenine dinucleotide (FAD); B, Flavine quinones and semiquinones.

Enzyme-FAD (oxidized) + AH ⟶ Enzyme-FADH (reduced) + A (4)

Many important compounds are found as substrates, including NADH (coenzyme), a-hydroxy acids, aldehydes, saturated hydrocarbon chains, some dyes (such as methylene blue), and even oxygen. The mitochondrial electron transport chain plays a critical role in providing energy to the cells of higher organisms. The formation of flavin semi-quinones during this process is an important step.

Because there are so many metabolic side reactions, no all possible flavine semi-quinones are known. For example, 40 years ago, the Nobel Prize winning chemist A Szent Gyorgyi discovered a reaction between riboflavin (vitamin B_2) and serotonin (a neurotransmitter). During this reaction, significant flavine semiquinones and several transfer complex semiquinones are formed. Later it was discovered that riboflavin reacts similarly with hallucinogens such a lysergic acid and benzyl dimethyl bufotenine. No biological significance of these reactions are known.

Because most flavine enzymes possess two molecules of flavin and each of these has at least three possible reduction paths, there is enormous complexity in these reactions. The importance of flavin containing enzymes is emphasized in table 2.1. Only the most important enzymes are listed in this table. These possess a high molecular weight and complex structure. Most of them are involved in mitochondrial respiration.

While dehydrogenases catalyze one electron transfer reactions, oxidases are involved in two step reactions, with each step transferring one electron. The electron polarization of red flavine semi-quinone (the product of oxidases) allows for significant stability. Dehydrogenases produce the end products of water or peroxide. Oxidases, on the other hand, form significant amounts of reactive oxygen species (ROS).

Flavin-enzymes possess a low specificity towards substrates, so that oxidases can act in many reactions, involving purines, aldehydes, dyes, etc. For example, Schardinger discovered, in 1902, that xanthine oxidase from milk is able to catalyze the bleaching of methylene blue by formic aldehyde. It is now known that xanthene oxidase is able to act on at least 60 neutral and synthetic compounds.

Xanthene oxidase (xanthene: oxygen oxidoreductase E.C.1.2.3.2) is classified as one of the most complex enzymes. As shown in table 2.1, it contains two FAD, eight iron atoms, and molybdenum and has a molecular weight of 300,000. In cells it is usually found as xanthene dehydrogenase, but is converted to this dangerous oxidative form during ischemic conditions. Xanthene oxidase can accept up to 14 electrons from varied substrates resulting in the formation of three types of flavin semi-quinone free radicals. It is accepted that an internalized electron transfer cycle must exist within the xanthine oxidase molecule. As will be shown in a future chapter, xanthene oxidase seems to be an important source of free radicals in some pathological conditions that involve ischemia.

It should be mentioned that aldehyde oxidase (E.C.1.2.3) is very similar in structure and properties to xanthene oxidase. Scientists from Wellcome Laboratory (Triangle Park, North Carolina, USA) compared the physiological levels of both enzymes on a phylogenetic scale from algae to human. They found that both enzymes shifted in affinity from NAD > ferricyanide > oxygen in fish, reptiles, and birds to ferricyanide > oxygen > NAD in monkeys and humans. Herbivores have a higher level of the enzyme than carnivores and humans are included in the later group. The evolution of the enzyme affinities for certain substrates may be correlated with nutrition type. This study might also illustrate the adaptability of enzymes at the molecular level [54, 154].

The great reactivity of semi-quinone radicals assures an increased affinity of enzymes and a rapid electron transport. The great complexity of flavin enzymes is advantageous for organisms, ensuring their activity in many reactions.

Table 2.1 The main flavin-enzymes and some of their properties.

Enzyme	Molecular weight	Coenzyme & active components	Physiological acceptor
L-aminoacid oxidase	138,000	2 FAD	Oxygen
Acyl CoA-dehydrogenase	200,000	2 FAD	Respiratory chain
Lipoyl dehydrogenase	100,000	2 FAD, -S-S-	NAD
NADPH cytochrome b5 reductase	40,000	FAD, Mg^{2+}	Cytochrome b
NADPH cytochrome c reductase	70,000	FAD	Cytochrome c
NADPH quinone reductase	60,000	FAD	Quinone (vit. K)
NADPH glutathione reductase	80,000	FAD	Oxidized glutathione
Xanthine oxidase	300,000	2 FAD, 8 Fe, 3 Mo	Oxygen, NAD, Purines
Aldehyde dehydrogenase	280,000	2 FAD, 8 Fe, 2 Mo	Oxygen, Aldehydes
NADH dehydrogenase	250,000	FMN, 8 Fe	Respiratory chain
Succinate dehydrogenase	300,000	FAD, 8 Fe	Respiratory chain

Table 2.2 The effects of ionizing radiation on DNA [197].

Effect	Radiolytic yield (mmoles per liter per 10 Gy)
Splitting a chemical bond	0.4 to 10
Splitting a double bond	0.12 to 0.14
Splitting a hydrogen bond	6.6 to 60
Polymerization	0.08
Nucleotide base degradation	1.7
Release of base of chemical group	0.2

Table 2.3 Different roles of nitric oxide.

Tissue	Positive effect	Toxic effect
Blood vessels	Antiatherosclerotic effects	Septic shock, inflammation
Heart	Negative inotropic ischemia	Myocardial "stunning," septic shock
Lung	Ventilation-perfusion matching, mucus secretion	Immune complex-induced alveolitis
Kidney	Tubuloglomerular feedback, renin secretion	Glomerulonephritis, acute kidney failure
Central nervous system	Synaptogenesis, memory formation, cerebral blood flow, neuroendocrine secretion, olfactory system	Neurotoxicity, convulsions, migrain, hyperalgesia
Pancreas	Endocrine & exocrine secretion	Cell destruction
Gut	Blood flow, peristalsis, exocrine secretion, mucosal protection, antimicrobial effects	Mutagenesis, mucosal damage
Immune system	Antimicrobial, possibly antitumor effects	Graft rejection, inflammation, septic shock, tissue damage

2.2 QUINONES AND SEMIQUINONES

Semiquinones are more widespread than flavin semiquinones. Quinones are very widespread, being found in respiratory chains, photosynthesis, as drugs, dyes, atmospheric pollutants, and cigarette smoke. Semi-quinones are characterized by a relatively high stability towards oxygen due to the electronic conjugation found in the aromatic ring.

Quinones provide a simple and effective reactive system that is widely used in electron transport. The conjugation of quinones offer excellent conditions for the addition of compounds, or charge transfer. Quinones are able to transport one or two electrons, or two electrons in successive steps, forming a semi-quinone as an intermediate. This two step process is commonly used in the electron transport chain, which is verified by the strong ESR signals characteristic of semiquinones found during mitochondrial respiration.

The transfer of one electron to a quinone usually forms a semi-quinone. This is observed in flavin enzymes such as NADPH-cytochrome P_{450} reductase or NADH ubiquinone oxireductase. The interaction of a semiquinone with oxygen results in the reoxidation of the quinone and the formation of superoxide ($O_2^{\cdot}$), a free radical. This occurs, with one electron transferred, during the metabolism of the insecticide paraquat. Vitamin K_3 reduction occurs with the transfer of two electrons. However, vitamin K_3 is a natural compound and the body is equipped to metabolize it without free radical formation through the diaphorase pathway.

Semi-quinones are able to rapidly cross cell membranes before reacting with oxygen to produce the free radical superoxide. At a concentration of 1µM semi-quinone and 250 µM oxygen, the resulting superoxide concentration will be about 0.7 µM, easily enough to produce adverse effects [8, 167].

In addition to their role in oxidation-reduction reactions, semi-quinones have had an important role in the history of science. Michaelis' studies of free radicals started with the oxidation of dimethyl p-phenylene-diamines (fig. 2.3). These compounds are used in the dye industry. Michaelis noticed that bromine water (a soluble form of bromine) oxidizes phenyl-diamines producing an array of yellow and red colors. The final product is called Wurster red, which is a long lived, stable free radical. Closed structure compounds are also produced from human plasma, peroxidases from plants, and enzymes from liver microsomes. Consequently, phenylene-diamines are now used to measure ceruloplasmin activity or mitochondrial electron transport activity.

Another important group of semiquinones are those produced during metabolism of cytostatic anthracycline drugs (adriamycin). These drugs act *in vivo* by producing semiquinones and superoxide. The semi-quinones bind to DNA, blocking cell proliferation. More about these free radical producers will be discussed in chapter 3.

The final interesting group of semi-quinones belong to a group of wide spread compounds such as polyphenols. Polyphenols are easily oxidized by peroxidases found in plants. When many plants are infected with parasites or viruses, peroxidase activation occurs, triggering the oxidation and polymerization of phenols. This process releases powerful reactive oxygen substances (ROS, oxygen free radicals) that are used as killing agent against microbes. Chemiluminescence has been demonstrated during this reaction [72, 123]. A similar killing system is found in phagocytotic leukocytes [191].

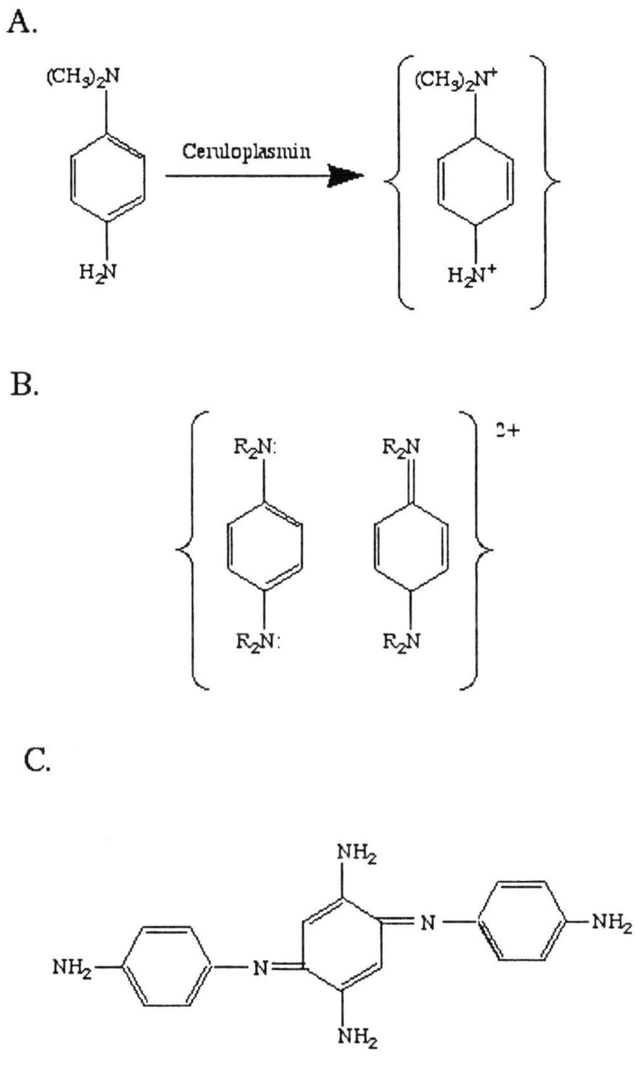

Figure 2.3 Free radicals during the oxidation of phenylene diamine by ceruloplasmin. A, The probable structure of the free radical; B, The possible structure of the Bandrowski base; C, Possible final product.

2.3 AROMATIC COMPOUNDS

Many aromatic compounds may be coupled with reactions that produce cationic free radicals such as $HOOC-C_6H_5-\dot{N}H$. A large array of aromatic compounds, such as aniline, p-aminobenzoic acid, aminopyrine, tertiary amines, etc. Are able to enter the organism as chemical pollutants and are then metabolized in the liver [78, 127]. Multiple oxidase system cytochrome P_{450} dependent or peroxidases are involved in this, complex, metabolism. The peroxidase pathway is nonspecific and, in general, consists of the following reactions (A = aromatic compound, C = intermediate compound).

H_2O_2 —peroxidase→ C-I
AH_2 + C-I —peroxidase→ C-II + $AH^{\bullet}$
C-II + AH_2 ⟶ free peroxidase enzyme + $AH^{\bullet}$

Therefore, metabolism of aromatic compounds by the peroxidase catalyzed pathway produces free radical aromatic compounds. It is for this reason the physiologically preferred pathway is the cytochrome P_{450} dependent pathway. The peroxidase system does exist in the organism and is probably used mainly when a the organism is presented with a large load of aromatic compounds to metabolize.

In a similar manner, phenols (R-OH) produce the phenoxy radical (R-$O^{\bullet}$). As will be discussed more in chapter 5.1, phenoxy radicals have great affinity for hemoglobin, resulting in the formation of methemoglobin. The peroxidase pathway for phenol is shown here (HbO_2 = Oxyhemoglobin, metHb = methemoglobin).

C-I + R-OH —peroxidase→ C-II + R-$O^{\bullet}$
HbO_2 + R-OH ⟶ metHb + R-$O^{\bullet}$ + H_2O_2

Peroxidase-catalyzed reactions are also involved in:

- Transformation of morphine into amorphine
- Metabolism of natural and synthetic estrogens
- Metabolism of polycyclic hydrocarbons, some of which are highly carcinogenic

Another group of aromatic compounds that produce free radicals are the 5-nitrofurans (which are antibacterials), 5-nitrotriazols, and 5-nitroimidazols. These aromatic drugs exert their cytolytic activity after being metabolized, when nitroreductases reduce the nitro group to amine. Unfortunately, the reduction of the nitro group favors the formation of free radicals with a high affinity for DNA, which can lead to mutations and cancer [36, 104]. The best know drug from this group is metronidazole (Flagyl) which is used to treat Giardia infections and as a radiosensitizer [116, 150]. There are multiple pathways possible in the metabolism of nitroderivatives (such as metronidazole and chloramphenicol) that produce free radicals having affinity for proteins, nucleic acids, or thiols, interfering with electron transport, or producing aromatic amines. Most of these compounds are harmful to the organism.

When nitrofurans are used as bacteriocidal agents, the formation of aromatic free radicals follows the action of intestinal bacterial nitroreductase. Thus, the aromatic free radical $RN^{\bullet}O_2^{-}$, is actually the bacteriocidal compound, but it can rapidly decompose by the reaction, $2\ RN^{\bullet}O_2^{-} \longrightarrow R-NO_2 + R-N=O$, which decreases the potency of the drug.

All nitroderivatives, when acting on an organism, interfere with oxygen with the consequences depending on the redox potential and the glucose concentration. The stimulation of oxygen consumption [20] can take place inside of cells or in their close vicinity. This property is useful when metronidazole and misonidasol are use as radiosensitizers in therapeutic settings. It is the goal in this case to produce a cytotoxic effect using these compounds. Ascorbic acid is added to the prescription to keep the radiosensitizers in the reduced, active, state.

An opposite effect is produced from aromatic free radicals produced from nitroderivatives such as 4-nitroquinoline-N-oxide and 4-hydroxyl-amino-quinolin-N-oxide. These aromatic free radicals are more likely to cause cancer.

The metabolism of aromatic compounds is always complex. Several reactions can occur resulting in varied products that possess different biological properties. For example, consider two closely related drugs, such as the antiepileptic clonazepam (5-chlorophenyl-7-benzodiazepin-2-one) and the coronary dilator nifendipin (2,6-dimethyl-2-nitrophenyl-3,5-piridin carboxylic acid). While clonazepam is metabolized to produce a free radical ESR signal, nifedipin does not. Clonazepam is metabolized in liver microsomes by the NADPH-cytochrome P_{450} reductasc (a flavin enzyme). Nitrofurantoin is a nitroderivative drug with great antimicrobial activity that is recommended for treatment of renal infections. Nitrofurantoin produces an aromatic free radical and superoxide radical, which are responsible for the side effects of pyelonephritis, pulmonary edema, and anemia. The same side effects can occur after administering another antimicrobial, chloramphenicol, or after accidental administration of the herbicide paraquat [69, 74, 116]. The use of paraquat as an herbicide is based on its capacity to produce free radicals that destroy the chloroplasts of plants. The paraquat free radical is known as dipirydilium and is produced even in anaerobic conditions, although without the superoxide produced in aerobic conditions.

2.4 NUCLEIC ACIDS

Free radicals produce from nucleic acids are the most interesting and useful in understanding the biological consequences of free radicals. Nucleic acid free radicals were discovered 40 years ago, but their study is not yet complete. Since nucleic acids are the basis of genetic information and inheritance, the behavior of their free radicals help explain some biological issues like mutation.

Eisinger and Shulman were the first to irradiate DNA samples *in vitro*, in the 1950s, and noticed the production of free radicals. This experiment opened an extensive study of radiation effects on nucleic acids and the consequence of free radical formation. Significant ESR signals were obtained for several of the nucleic acids, and especially with purine and pyrimidine bases.

All experiments concerning the formation of free radicals are complex, and those involving nucleic acids are even more so. Following the absorption of radiation energy by nucleic acids, many modifications occur within a fraction of a second. Most of these modifications are ionizations and excitations. If the excited molecule or atom then loses the absorbed energy in fractions of a second, long lasting free radicals are produced. Therefore, nucleic acid free radicals are formed following the absorption of energy (ionizing or UV) leading to splitting and other chemical modifications (table 2.1).

Thymidine, a pyrimidine base, is the most affected of the nucleic acids. A hyperfine structure in ESR signals is observed due to the electronic interaction. The radiolytic yield of primary free radicals in irradiation sickness is 6×10^{-6} Moles/liter per Gy (1000 Rads), which is a lethal dose for all mammals. The ESR signals from irradiated tissues are more complex because water molecules interact by binding with the nucleic acid as well as with metal ions present. ESR signals have also been obtained from other nucleic acids, such as S-methyl-cytosine, cytosine, 5-hydroxyuracil, guanine, and xanthine.

Signals indicating the presence of the hydroxyl radical (OH˙) have also been obtained. The hydroxyl radical results from the irradiation of water in the cells (water radiolysis). This reactive oxygen specie is the most important free radical produced by radiation. Hydroxyl radical can destroy any organic compound, including nucleic acids, which they attack at a very high rate ($k = 7 \times 10^9 \, M^{-1}S^{-1}$). Radiobiological studies have shown that some protection may be provided DNA with sulphydryl compounds, such as cysteamine. If in close proximity to the DNA, it may provide protection by acting as the target for the reactive oxygen species.

The importance of nucleic acid free radicals is immense. As will be discussed later in greater detail, the biological consequences of oxidative stress (chapter 5) and radiation sickness (chapter 6) are largely due to the formation of free radicals and other reactive oxygen species and the stable compounds resulting from the degradation of nucleic acids. These degradation products were first discovered in the urine of animals exposed to experimental radiation and of men accidently exposed to high levels of ionizing radiation.

Recent biochemical studies have emphasized the importance of DNA-protein complexes in the study of gene control mechanisms (a technique called fingerprinting). In 1988, Tullis devised an new variant of DNA fingerprinting. It exposes only a portion of the DNA molecule (that which is not involved in protein binding) to a free radical generating system (iron salt and EDTA). After separation, the degraded DNA is isolated with high resolution electrophoresis providing a map of the DNA.

The free radical generating system functions according to the following equations.

$$FeCl_2 + EDTA \longrightarrow [Fe(EDTA)]^{-2}$$
$$[Fe(EDTA)]^{-2} + H2O2 \longrightarrow [Fe(EDTA)]^{-1} + OH^- + OH^\bullet$$

The actual reactive oxygen species (ROS) formed is the hydroxyl radical (OH˙).

2.5 THYLIL RADICAL

The existence of the thylil radical (R-S˙) was disputed for a long time because it cannot be detected by ESR or spectroscopy. As was the case for other free radicals, the thylil radical was theoretically predicted from the analysis of ESR signals from *in vitro* irradiated amino acids and proteins [197, 198]. Sulphur containing compounds may be simple thiols, such as cysteine or glutathione, or the free SH groups found in proteins. The presence of simple thiols protects nucleic acids and proteins against reactive oxygen species attack. This is because thiols and SH groups in proteins are the favored target of free radicals, even for those with a short half-life such as are produced during enzymatic activity. Therefore, the free SH groups are located in the active center of many enzymes or hormones. During catalytic activity, a free thyil radical can be formed as the result of a transfer reaction.

Protein-SH + X$^{\bullet}$ ⟶ R-S$^{\bullet}$ +XH

The thyil free radical has a very short life (less than 10^{-5} sec.). It was only in 1990 that Maples, Euer, and Mason, working at Triangle Park, North Carolina, USA were able to indirectly detect R-S$^{\bullet}$ by using ESR spin trapping in a reaction between free SH groups of hemoglobin and a nitro-aromatic compound (nitrosobenzene) [112].

2.6 NITRIC OXIDE

Nitric oxide is a short lived gaseous free radical. It has long been known to exist in the atmosphere, and is now known to also be a very important biological compound [145, 196]. There is now a great deal of experimental evidence that nitric oxide (NO) is a potent multifunctional metabolite that can act as a neurotransmitter, a vasodilator, can contribute to immunosuppression, to blocking cell adhesion, and may serve as a host defense molecule.

Even in 1916, Mitchael suggested that humans may synthesize oxides of nitrogen. Nitroglycerine has been used for over a century to treat coronary heart disease, so it should be no surprise that endogenous NO affects blood vessels and circulation. It is now known that nitric oxide is endothelium derived relaxing factor (EDRF). Nitric oxide derives from several cellular sources. Nitric oxide is formed in endothelial and epithelial cells. Neurons use NO to regulate transmitter release of adjacent neurons and to regulate cerebral blood flow. Bronchial epithelial and endothelial cells use NO to match ventilation and perfusion.

Nitric oxide is, however, a double edged sword. It plays a role as messenger or modulator for immunologic defense, but NO may also be toxic, generating reactive oxygen species and decreasing the effectiveness of antioxidant systems.

$$O_2 + NO \longrightarrow ONOO^- \text{ (peroxynitrite)} \xrightarrow{H^+} ONOOH \longrightarrow OH^{\bullet} + NO_2^- \longrightarrow NO_3^- + H^+$$

Peroxynitrate is a reactive oxygen specie that is cytotoxic, which destroys tyrosine residues in proteins.

Nitric oxide is able to inhibit platelet aggregation and agonist-evoked cytoskeletal reorganization of the fibrinogen receptor glycoprotein. The possible roles of NO are presented in table 2.3. As seen in the table, NO can exhibit opposite actions on the same tissue.

Nitric oxide possesses an extra electron, making it highly reactive with a variety of different molecular targets, especially proteins. This interaction leads to functional modifications. NO is synthesized from L-arginine by the enzyme NO synthase.

Arginine —Synthase→ Citrulline + NO

Nitric oxide possess a half-life of seconds, but, due to its highly reactive nature, it interacts with several compounds, such as proteins, thiols, and hemes. The biochemistry and medical applications of NO are far from being entirely known.

Chapter 3

OXYGEN FREE RADICALS

3.1 THE OXYGEN PARADOX

The free radicals discussed in the previous chapter are formed from various organic compounds, most of which possess an aromatic structure. But these free radicals are actually of limited importance because of their short life time and restricted formation. Their main importance is in their role in the initiation of the formation of activated oxygen.

Oxygen plays an essential role in the chemistry of free radicals. First, oxygen is present everywhere in the atmosphere, in all organisms, and in most compounds found on the earth. Table 3.1 lists the amount of oxygen in some of these locations. Therefore, the interaction of oxygen with free radicals produced in the types of reactions described in the previous chapter is almost inevitable.

Table 3.1 The oxygen content of the atmosphere and organisms

Sun's surface		4.0%
Earth	- lithosphere	46.6%
	- hydrosphere	85.8%
	- atmosphere	23.0%
Human body	- wet weight	65.0%
Air	- inhaled	21.0%
	- exhaled	14.4%
	- alveolar	13.8%

Secondly, oxygen's electron structure makes the molecule especially susceptible to free radical formation. As illustrated in figure 3.1, oxygen possesses two nonpaired electrons, in separate orbitals, in its outer electron shell. However, these unpaired electrons have the same spin, making the molecule a low reactive radical. Pure free oxygen is restricted in oxidative reactions, but the presence of catalytic metals (Cu, Fe, Mn), electron donors, or free radical-initiating reactions results in the activation of oxygen. The activation of oxygen means its electron structure is modified through spin inversion, making a highly reactive formation. Molecular oxygen has a short bond length of 1.12 Å. Following its activation, the molecular bond length increases to 1.33 Å for superoxide ($O_2^·$) and 1.49 Å for the peroxi radical (O_2^{2-}). The binding energy decreases from 118 to 64 to 51 kcal/mol. These modifications result in a very reactive molecule that is easily able to participate in oxidative reactions.

Oxygen's appearance in the atmosphere, some 3 billion years ago, was the result of volcanic activity, photochemical reactions in the atmosphere. The early forms of life were unicellular anaerobic organisms that eventually developed the capacity for photosynthesis. For these early organisms, oxygen was a waste product that they had to develop ways of protecting against. The result was the first antioxidants. As the oxygen content in the atmosphere increased, those organisms with efficient antioxidant systems were able to survive and eventually learn to use the paradoxic nature of oxygen. The use of oxygen involved the formation of oxidoreductive enzymes and lead to the development of mitochondrial respiration and all the other parts of the aerobic higher organism.

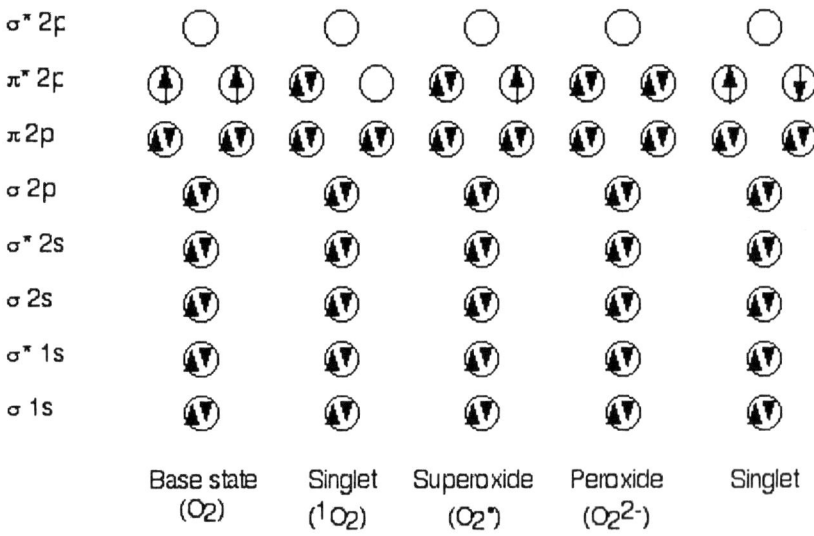

Figure 3.1. The electron structure of oxygen and the spin restrictions of various forms.

These processes generally used enzymes that contained metallic ions in their active centers. These metals (Cu, Fe, Zn, Mn) in the free state are able to activate oxygen, but, when bound in a protein, catalyze oxidoreductive reactions of benefit to organisms, including the decomposition of reactive oxygen species.

All this helps define the oxygen paradox. On the one hand, oxygen favors higher forms of life. The pathways that evolved to handle oxygen provide greater amounts of energy than is possible from anaerobic pathways. On the other hand, oxygen is very toxic towards life.

Depending on the electronic configuration (fig. 3.1), more or less stable free radicals are formed. Various scientific congresses have decided to use the term reactive oxygen species (or ROS) for these rather than oxygen free radical [22, 129, 140, 167]. As will be seen, oxygen has various reactive forms. Some of these are free radicals, such as superoxide ($O_2^{\cdot}$) or the hydroxyl radical ($OH^{\cdot}$). However non-free radical forms also exist, such as the electronically excited form called singlet oxygen ($^1O_2$), and the stable compound hydrogen peroxide (H_2O_2).

Reactive oxygen species appear during a basic process: the reduction of oxygen to water, as occurs during respiration. This process primarily takes place in the mitochondria, but also occurs in many enzymatic catalyzed reactions. As discussed in chapter 2 (see reactions 1 and 2), oxidases catalyze oxidoreductions of 2 and 4 equivalents and no ROS appear. The reduction of oxygen to water requires four electrons, which can be transferred in successive one electron steps. When this occurs, toxic oxygen species are formed, beginning with superoxide. This radical has little damaging capacity, but it is transformed into more deadly forms, terminating in the hydroxyl radical.

Life is organized to favor the tetravalent reduction of oxygen to water, preventing the formation of ROS. However, univalent reduction of oxygen does occur in the body and amounts to about 5% of the total oxygen consumption. The elimination of ROS formed in this process is the purpose of the physiological antioxidant systems.

3.2 SINGLET OXYGEN

As shown in figure 3.1, oxygen, in order to avoid spin restrictions, can exist in an electronically excited form called singlet oxygen, which is reactive enough to damage tissue (22 kcal/mol). Singlet oxygen is formed following absorption of energy quanta resulting in an electron jump to a superior orbital. Its life is very short (thousands of a second), but still long enough to produce damaging effects. The presence of singlet oxygen has been demonstrated with chemiluminescence and illumination of chloroplasts, in chlorophyll, flavins, porphyrins, and the retina [172]. Singlet oxygen reacts with some organic compounds, such as polyunsaturated fatty acids (PUFA), cholesterol, and hydrocarbons, forming peroxides or structurally modification. In 1990, Krinsky, at Loyola University, detected singlet oxygen in human plasma using a phosphorescence spectrometer coupled to a dye laser [99]. He found that singlet oxygen in human plasma has a life span of 1.04 ± 0.03 µs. Superoxide that appears in plasma is quickly decomposed or quenched by albumin or other antioxidants present in plasma. Antioxidants in plasma capable of quenching superoxide include vitamins C and E, bilirubin, and carotene.

A poorly understood reaction is that of singlet oxygen with tryptophan. This reaction seems to be involved in photosensitizing reactions, especially in proteins in the eye (crystalline). As will be discussed in chapter 6.3, the formation of singlet oxygen may be involved in the formation of cataracts (crystalline is rich in tryptophan).

Singlet oxygen has also involved in the etiology of some drug-induced dermatitis produced by cosmetic, plant toxins, or porphyrins [92, 114]. In all these conditions, singlet oxygen is formed in photosensitizing reactions triggered by metabolites of these compounds.

Another side of singlet oxygen seems to be its possible implications in phototherapy. The treatment of psoriasis with psoralens or treatment of pulmonary carcinoma with hematoporphyrin results in the formation of singlet oxygen. The singlet oxygen formed seems to have a primary effect on diseased cells. Unfortunately, this explanation does not always hold up. Psoriasis patients who eat a diet rich in carotenoids may improve with treatment, even though carotenoids are good quenchers of singlet oxygen [25, 107, 169].

3.3 SUPEROXIDE

Superoxide radical (see fig. 3.1) is the first reactive oxygen specie formed during the univalent reduction of oxygen. Superoxide is formed when oxygen undergoes electron capture and its localization on an antibonding orbital. Thus, superoxide is an anionic free radical, explaining the basis for its notation as $O_2^{\bullet}$ or $O_2^{\bullet-}$.

As is oxygen, superoxide is paramagnetic and exhibits an ESR profile. The ESR signal is strongly dependent on temperature. In media like acetonitrile or dimethyl sulfoxide, superoxide is a powerful reducing and nucleophilic agent [2, 99, 114].

Superoxide can be produced relatively easily in the laboratory through irradiation of water with gamma rays or by pulse radiolysis. Superoxide results during the interaction of water radiolysis products with oxygen.

$$H_2O \longrightarrow e^-_{aq} + OH^{\bullet} + H_2O + H^{\bullet} \tag{1}$$

$$e^-_{aq} + H_2O \longrightarrow O_2^{\bullet} \tag{2a}$$

$$H^{\bullet} + O_2^{\bullet} \longrightarrow HO_2 \tag{2b}$$

Superoxide is also produced during exposure of hydrogen peroxide to UV radiation or during *in vitro* oxidation of flavins, polyphenols, and quinones. There is indirect evidence that superoxide can be produced in vivo (table 3.2).

A summary of superoxide reactions (dismutation, Habe-Weiss, Fenton) is now presented.

Dismutation reactions of superoxide are central to its action. The dismutase reaction consists of an electron transfer from one molecule of superoxide to another. This dismutation results in the production of hydrogenperoxide and oxygen. However, there are three possible dismuation reactions (note: $HO_2^{\bullet}$ is the protonated form of $O_2^{\bullet}$).

Table 3.2 Sources of superoxide in living organisms

Endogenous sources	Electron transport chain in mitochondria, chloroplasts, and endoplasmic reticulum
	Oxidant enzymes: xanthine oxidase, galactose oxidase, monoamine oxidase, tryptophan oxigenase, cyclooxygenase, lipoxidase
	Phagocytic cells: neutrophiles, monocytes, macrophages, eosinophiles, endothelial cells.
Exogenous sources	Autooxidation of catecholamines, metabolic ferrous ion, thiols, hemoglobin
	Oxidation of redox compounds: paraquat, anthracycline drugs, alloxan
	Metabolic oxidation of paracetamol, carbon tetrachloride, etc.
	Irradiation with ionizing radiation, UV light
	Burns and other trauma

$$HO_2^{\bullet} + HO_2^{\bullet} \longrightarrow H_2O_2 + O_2 \qquad K2 = 7.6 \times 10^5 \, M^{-1}s^{-1} \qquad (3)$$

$$HO_2^{\bullet} + O_2^{\bullet} + H^+ \longrightarrow H_2O_2 + O_2 \qquad K2 = 8.5 \times 10^7 \, M^{-1}s^{-1} \qquad (4)$$

$$O_2^{\bullet} + O_2^{\bullet} + 2H^+ \longrightarrow H_2O_2 + O_2 \qquad K2 = 6.0 \, M^{-1}s^{-1} \qquad (5)$$

These reactions are of greatest importance to living organisms. Reaction (3) proceeds very rapidly at pH 4.8, which does not occur in living organisms. Consequently, in vivo, the dismutation reaction only occurs with the aid of the enzyme superoxide dismutase. It has been demonstrated that oxygen was adapted by early forms of life only after the ancient anaerobic cells developed an antioxidative system consisting of superoxide dismutase (SOD) [2, 123, 136].

It is apparent that ancient forms of life experimented to develop an efficient enzyme by using different metals, such as copper, zinc, manganese, and iron, as the active center. Superoxide dismutase catalyzes reaction (5), which proceeds spontaneously only very slowly. In the presence of the enzyme, the reaction proceeds rapidly ($10^9 \, M^{-1}s^{-1}$). The importance of the SOD-catalyzed reaction is demonstrated by the presence of the enzyme in aerobic cells. Cells exposed continuously to a high level of oxygen also contain large amounts of superoxide dismutase. Likewise, anaerobic cells are induced to produce superoxide dismutase when exposed to oxygen.

As stated earlier, several metals are used as an active center for superoxide dismutase.

$$O_2^{\bullet} + M^{n+} \longrightarrow M^{n-1} + O_2 \qquad (6a)$$
$$O_2^{\bullet} + M^{n-1} + 2H^+ \longrightarrow M^{n+} + H_2O_2 \qquad (6b)$$

Another important reaction of superoxides is the Haber-Weiss reaction, discovered by Haber and Weiss in 1934. This reaction occurs between superoxide and hydrogenperoxide.

$$O_2^{\bullet} + H_2O_2 \longrightarrow O_2 + OH^- + OH^{\bullet} \qquad (7)$$

It results in a mixture of oxygen and the hydroxyl free radical. The reaction takes place very easily *in vitro* in a variant known as the Fenton reaction.

$$Fe^{2+} + H_2O_2 \longrightarrow Fe^{3+} + OH^- + OH^{\bullet} \tag{8}$$

The importance of the Fenton reaction consists in the formation of the hydroxyl radical (OH˙), which is the most powerful of the reactive oxygen species. This free radical is capable of destroying any organic molecule and living systems have no effective defense against it.

The existence of the Habe-Weiss or Fenton reaction *in vivo* is a hot subject in the biochemistry of free radicals. Most books dealing with free radicals mentions this problem and presents an opinion. There is no clear cut evidence about this, and the pro and con positions are all based in *in vitro* experiments under conditions that approach *in vivo* conditions.

The Haber-Weiss reaction has several variants. Thus, reaction (8) also produces a short-lived chemiluminescence that proves the production of superoxide. Also, copper salts react more rapidly than ferrous ions.

$$Cu + H_2O_2 \longrightarrow Cu^{2+} + OH^- + OH^{\bullet} \tag{9}$$

While superoxide is eliminated by superoxide dismutase, the formation of the hydroxyl radical is inhibited by several compounds, such as mannitol, formate, thiourea, and the chelating drug desferrioxamine.

The anion, superoxide, may be involved in several electron transfer reactions with acceptor compounds such as:

Quinones: $quinones + O_2^{\bullet-} \longrightarrow semiquinones + O_2$ (10)

Bromouridine: $BrU + O_2^{\bullet-} \longrightarrow U^{\bullet} + Br^- + O_2$ (11)

Thiols: $R\text{-}SH + H^+ + O_2^{\bullet-} \longrightarrow RS^{\bullet} + H_2O_2$ (12)

In all these reactions, superoxide triggers other free radical producing processes.

It is clear that superoxide production occurs within the living organism during physiologic activity (table 3.2). It is for this reason intracellular antioxidant systems are so important.

3.4 HYDROGEN PEROXIDE

The addition of a second electron to superoxide produces the peroxide ion (O_2^{2-}), which is not a free radical. Under conditions existing in vivo, peroxide ion that is formed will produce hydrogen peroxide (H_2O_2). Hydrogen peroxide is formed in organisms in significant concentrations (10^{-3}M) in many reactions. Hydrogen peroxide is formed during flavin coenzyme catalyzed reactions, oxidation of quinones, oxidation of sulphydryl containing compounds, and superoxide dismutase activity. But, the most important source of hydrogen peroxide formation in cells is the electron transport chain in mitochondria. In this reaction, there is a continuous production of 82 nmoles of hydrogen peroxide per gram of wet tissue per minute (105). In mammalian eyes, micromolar concentrations of hydrogen peroxide have been measured, the same level as has been found in exhaled air. This is clear evidence of the continuous formation of hydrogen peroxide in living organisms as a result of physiologic activity.

Living organisms have evolved a mechanism to rid themselves of this compound. In fact there is more than one system in aerobic cells that are able to quickly and efficiently decompose hydrogen peroxide.

The decomposition of hydrogen peroxide involves a variant of the Fenton reaction. Iron-containing proteins were used by early life forms to decompose hydrogen peroxide. These iron-containing enzymes are catalase and peroxidases, which act as follows.

$$H_2O_2 + H_2O_2 \text{ --- catalase-Fe}^{3+} \rightarrow O_2 + 2\,H_2O \tag{13}$$

$$H_2O_2 + R\text{-}H_2 \text{ --- peroxidase-Fe}^{3+} \rightarrow R + 2\,H_2O \tag{14}$$

In contrast to the Haber-Weiss (7) and Fenton (8) reactions, which are harmful to life, reactions 13 and 14 are beneficial, producing oxygen and a metabolite (R).

Among reactive oxygen species, hydrogen peroxide is the single stable compound that is relatively easy to measure. Thus, its presence in living organisms was long ago demonstrated.

A second, difficult to explain, characteristic of hydrogen peroxide is the low reactivity of commercial preparations of the compound compared to the reactivity of the compound produced *in vivo*. It is suggested that when hydrogen peroxide is produced in tissues, the total presence of enzymes results in the production of hydroxyl radical together with hydrogen peroxide [85].

A third trait of hydrogen peroxide is its cytolytic nature. Depending on the intracellular content of catalase in a bacterium, these unicellular organisms may be killed by the continuous secretion of hydrogen peroxide [86]. A similar killing mechanism exists in phagocytizing leukocytes and other similar defensive cells. The sensitivity of microorganisms toward hydrogen peroxide is an accepted criterion for bacterial selection and is directly related to the cells content of catalase.

3.5 HYDROXYL RADICAL

Singlet oxygen ($^1O_2$) and the hydroxyl radical (OH˙) are the most powerful reactive species of oxygen.

Hydroxyl radical is easily produce in vivo by the decomposition of hydrogen peroxide when heated or by ionizing radiation. In vivo, the hydroxyl radical is produced in reaction (7, 8, and 9). It reacts readily with any organic compound (RH) such as nucleic acids, lipids, carbohydrates, and proteins.

$$OH^{\bullet} + RH \longrightarrow R^{\bullet} + H_2O \tag{15}$$

$R^{\bullet}$ is a free radical or structurally modified compound. It is possible that the great damaging capacity of the hydroxyl radical is due to the formation of these modified compounds, which are then unable to exhibit their normal function. Indeed, exposure of bacteria or viruses to a hydroxyl radical generating system abolishes their ability to reproduce and inactivates them.

The reactivity of hydroxyl radical exceeds that of the superoxide radical and is among the highest in all chemistry. Whether its action is direct or indirect, the formation of hydroxyl radical in close proximity to critical biological molecules, such as DNA or regulatory enzymes, is likely to have an oxidative effect. Close proximity on the cellular scale means approximately 20Å, or roughly 5 molecular diameters. The reaction of hydroxyl radical with organic molecules leads to the structural modification of that compound and the transitory formation of organic peroxides ($RO_2^{\bullet}$).

$$RH + OH^{\bullet} \longrightarrow R^{\bullet} + H_2O \longrightarrow O_2 \longrightarrow ROO^{\bullet} \tag{16}$$

Hydroxyl radical has been measured in these types of reactions using ESR.

Given this great reactivity, it is less of a wonder that no effective antioxidant system against the hydroxyl radical has evolved. Living organisms have had to content themselves with the strategy of defending themselves from the hydroxyl radical by removing its precursors, such as superoxide and hydrogen peroxide. Under laboratory conditions, the formation of the hydroxyl radical is completely prevented by compounds such as mannitol or thiourea, although they are destroyed in the process.

Antioxidant systems present in cells, that breakdown superoxide or hydrogen peroxide, avoid the production of hydroxyl radicals. Metallothioneins are tissue proteins that bind strongly to a large array of metallic ions. This removes the ions from being able to catalyze the formation of hydroxyl radical, making reactions (6, 8, and 9) unable to proceed. Metallothioneins do efficiently decompose hydroxyl radical ($k = 10^{10}$ $M^{-1}s^{-1}$), but their concentration in the blood and tissue is too low to be physiologically significant.

Data indicates the hydroxyl radical, once formed in the blood, is efficiently eliminated following its reaction with plasma albumin ($k = 10^{10}$ $M^{-1}s^{-1}$) and with glucose ($k = 10^9$ $M^{-1}s^{-1}$) [29, 71, 168]. Plasma albumin is present at a high concentration (35 - 50 g/l), and glucose has a physiological level of about 4.5mM. Therefore, these two blood components also play an important protective role [73, 76]. Fibrinogen, another protein present at high concentration in the blood (3 g/l) may also be effective in decomposing the hydroxyl radical.

3.6 PEROXIDES

Peroxides are the most extensively studied of all reactive oxygen species. There are more than 50 books, numerous scientific congresses, and hundreds of scientific papers dedicated to this subject [8, 72, 123, 167, 177, 203].

There are several reasons for the great interest in peroxides. First, peroxides and their degradation products are stable. Second, peroxides are widely found in nature and in manufacturing processes (plastics, rubber, dyes, fuels, etc.). Third, there is a lengthy history of research in peroxides, going back at least as far as the second world war. In fact, by the 1950s research into the causes and mechanisms of peroxidation in manufactured products had been nearly completed. These studies included the aging of plastics and rubber, and the rancidity of fats. These studies found practical solutions to prevent or delay peroxidation in manufactured products. It was learned that when plastics, rubber, meat, etc. are exposed to UV light, metallic ions, and heat, the oxygen activation process is accelerated and peroxidation results. The first steps in peroxidation involve the formation of reactive oxygen species such as superoxide and hydroxyl radical [123, 177].

Once studies on peroxides that had an economic impact were completed, interest in research on peroxides declined. Only in the mid 1960s did peroxidation research centered on the biological consequences begin, because of concern about the effects of ionizing radiation. These radiobiological studies discovered that radiolysis of water results after exposure to radiation. In tissues and blood, the final result of exposure to ionizing radiation is the formation of lipid peroxides, which are final products of oxygen activation.

Figure 3.2 illustrates the formation of lipid peroxides and some of their decomposition products. It shows the formation of short-lived free radicals as described earlier in this chapter. The propagation reaction generates more free radicals that become the focus of more peroxide formation or other degradative reactions. The steps shown in figure 3.2 may all be taking place simultaneously or successively, depending on the local presence of substrates and antioxidants and on other conditions.

Peroxides (ROO$^\cdot$) are formed in the tissues of organisms as a consequence of hydroxyl radical (OH$^\cdot$) production. There is both direct and indirect evidence that the hydroxyl radical is able to initiate peroxidation of lipids, nucleic acids, and aromatic compounds [9, 39, 112, 203]. Experimental evidence shows that singlet oxygen is also able to cause peroxidation of substrates such as polyunsaturated fatty acids, olefines, aromatic compounds, and amino acids (particularly tryptophan) [29. 67, 167, 180].

Peroxidation, once it has begun, follows the same steps as free radical formation (initiation, propagation, termination). Peroxidation, however, possesses an unique characteristic. Once formed, peroxides generate their own catalysts for propagation.

$$ROO^\cdot + RH \longrightarrow ROOH + R^\cdot \tag{17a}$$

$$ROO^\cdot + RSH \longrightarrow ROOH + RS^\cdot \tag{17b}$$

These catalysts consist of other peroxides or their degradation products (aldehydes). Once begun, peroxidation can proceed to expand and reach all susceptible structures in a cell (unsaturated fatty acids, nucleic acids, heme-proteins, steroids, etc.).

During propagation, hydroperoxides react with other compounds (reaction 17) or with each other.

$$R_1OOH + R_2OOH \longrightarrow R_1OOR_2 + O_2 \tag{18a}$$

$$R_1CHOO + R_2CHOO \longrightarrow R_1C=O + R_2CHOH + O_2 \tag{18b}$$

The most likely reaction is one that involves polyunsaturated fatty acids (PUFA). This type of reaction occurs *in vitro* (for example canned meats) and *in vivo*, and once started is difficult to stop as new peroxyradicals (ROO˙) are produced and branch reactions appear. In addition, the decomposition products (aldehydes and ketones) also participate in prolonging the chain reaction.

For a peroxidative process to begin, favorable conditions are needed. These include the presence of pro-oxidative compounds, lack of antioxidants, and external parameters such as heat, radiation, etc. Even *in vitro* (tissue homogenates, plastics, rubber), the peroxidation process is very complex with lag time, reaction time, and final products depending on other compounds present in the medium.

The lag time is a function of the time needed to initiate the formation of free radicals and exhaust antioxidants in the system. Some of these are shown in table 3.3. There are two important conclusions that should be drawn from table 3.3. First, the enzymatic systems are the more efficient in triggering or stopping peroxidation. Secondly, some compounds, such as heme-containing proteins and ascorbate favor peroxidation in low amounts (μM concentrations) and stop it at higher concentrations (mM), reacting with nonenzymatic antioxidants to produce other free radicals of lower reactivity. Therefore, it is clear that the formation of free radicals and peroxidation are processes that are difficult to predict. It is for this reason that so many experiments lead to varied results, occasionally the opposite of that expected.

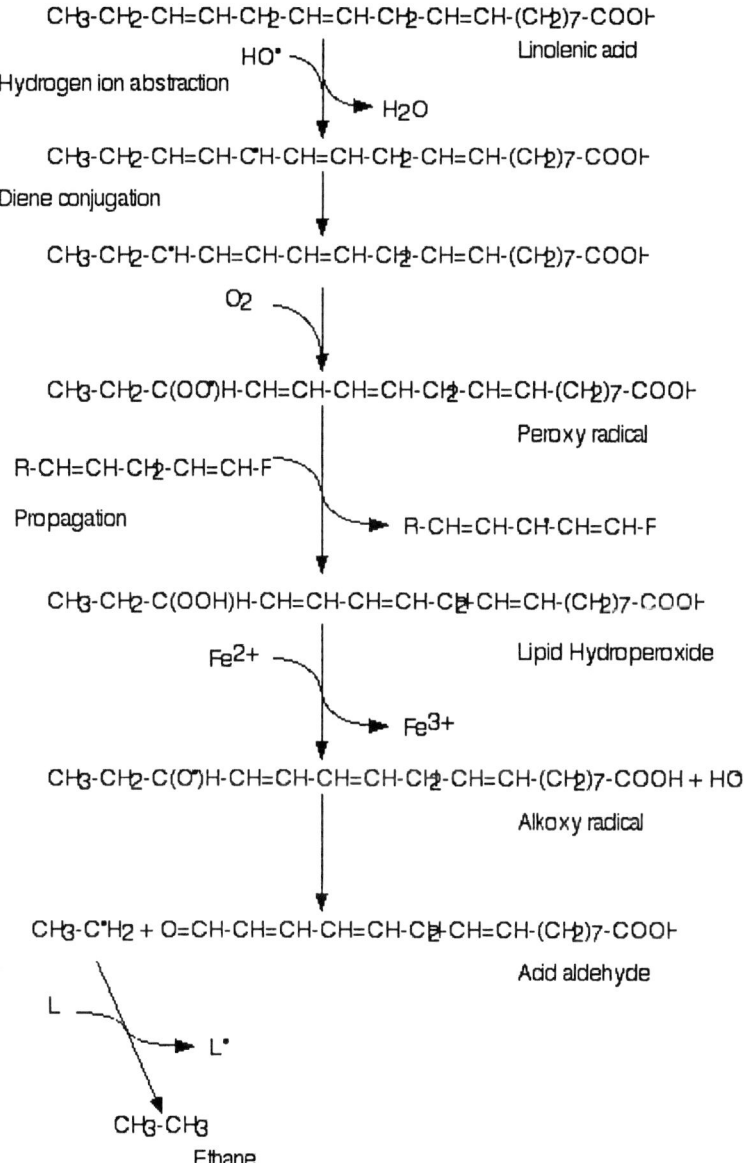

Figure 3.2. The main steps in the peroxidation and degradation of linoleic acid.

Table 3.3 The main natural pro- and antioxidants

Prooxidants	Antioxidants
Lipoxygenase	Superoxide dismutase
Cyclooxygenase	Glutathione peroxidase
Heme-containing compounds*	Heme-containing compounds*
Metalic ions (especially Fe, Cu, Co)	Chelating compounds
Ascorbate*	Ascorbate*
Quinones, aromatic metabolites	Albumin, uric acid melatonin, glutathione

*Depending on conditions and concentration.

The formation of peroxides in *in vitro* experiments using homogenates, liposomes, cells, etc. occurs under artificial conditions, where a trace of pre-existing peroxides or metallic ions may strongly influence the results [70, 140, 177, 92]. Therefore, it is difficult to extrapolate the formation of peroxides from *in vitro* experiments to the whole animal and to expect the detection of significant amounts of lipid peroxides in blood plasma. Never-the-less, lipid peroxides have been found by many scientists in the blood of healthy people by the use of numerous laboratory methods. This argues strongly for the physiological formation of peroxides [39, 124].

As shown in figure 3.2 and in reaction (18), a chemical group called a conjugated diene is produced during peroxidation. This structure consists of conjugated double bonds (R-CH=CH-CH=CH-R) having a light absorption maximum of 235 nm. This peak has been used to measure oxidized PUFA in lipids extracted from blood [9, 32, 180].

The enzyme, lipoxygenase (EC 1.13.11.12), should be mentioned at this point. Twenty years ago, this enzyme was only known in plants. Then it was found in animal cells (leukocytes and platelets) and in subcellular fractions (mitochondria, and testicular microsomes) [99]. This enzyme contains nonheme iron and catalyzes the peroxidation of PUFA into lipoperoxides with free radicals as intermediates. Cadenas [29] showed that this peroxidation is accompanied by chemiluminescence as some carbonyl compounds (R-C=O) are formed. Quantitative correlations were found between the intensity of the chemiluminescence, oxygen consumption, and the production of diene. Thus, the same enzyme forms peroxides and degrades them. This complicates the unraveling of peroxidative reactions in the living organism.

Because of this, it is not surprising the methods used to measure peroxidation are far from perfect. In the industrial laboratory (canned meats, plastics, etc.), chemical methods are used. But in biological samples, no direct methods are ideal, so indirect measurements are widely used. Chemical methods measure the degradation products of peroxides, often as malondialdehyde (MDA). Since it is not possible to measure free radicals in blood, the determination of MDA as an indicator of peroxidation is widespread. Another indicator that has been used is the isolation of nucleic acid degradation products from urine [5]. Recently, instruments have been able to measure the final decomposition products of peroxides (pentane and other hydrocarbons) in exhaled breath.

Chapter 4

NATURAL PROCESSES AS SOURCES OF FREE RADICALS

As stated at the beginning of chapter 3, the increased oxygen level in the earth's atmosphere 3 billion years age required the primitive forms of life to adapt by creating protective antioxidative systems. Those algae and bacteria that successfully adapted to the increased concentrations of oxygen began to use it as a promotor for redox processes and as an initiator for some essential processes.

4.1 MITOCHONDRIAL RESPIRATION

Respiration is a complex process. It culminates with oxygen from inhaled air reaching intracellular organelles called mitochondria. In eucariotic, aerobic cells, the final step of respiration consists of the reduction of oxygen by four electrons obtained from catabolic reactions and passed along the electron transport, respiratory chain. Until 20 years ago, this final step of respiration was believed to consist of the following reactions:

4 cytochrome c^{2+} + cytochrome oxidase (oxidized) →
4 cytochrome c^{3+} + cytochrome oxidase (reduced) (1)

O_2 + cytochrome oxidase (reduced) + $4H^+$ →
Cytochrome oxidase (oxidized) + $2H_2O$ (2)

This reaction does not produce free radicals as it is not an univalent reduction of oxygen.

In 1973, the work of B. Chance and A. Boveris of the Johnson Foundation in Philadelphia, Pennsylvania triggered a sensation among scientists by demonstrating the formation of significant amounts of superoxide and hydrogen peroxide during mitochondrial respiration [34, 100].

The electron transport chain consists of a chain of heterogenous redox catalysts such as nucleotides (NADH), flavoproteins (NADH dehydrogenases), and iron and sulfur containing proteins localized in the internal membrane of the mitochondria (fig. 4.1).

These proteins have a range of redox potentials. The sequence of reactions that take place along the respiratory chain permit the passing of electrons from NADH to O_2 with the reduction of 1/2 O_2 producing one H_2O. This chain couples with oxidative phosphorylation and produces ATP. As the rate of oxidative phosphorylation is the limiting step, any compound that decouples the biosynthesis of ATP will greatly decrease the oxygen consumption of the system.

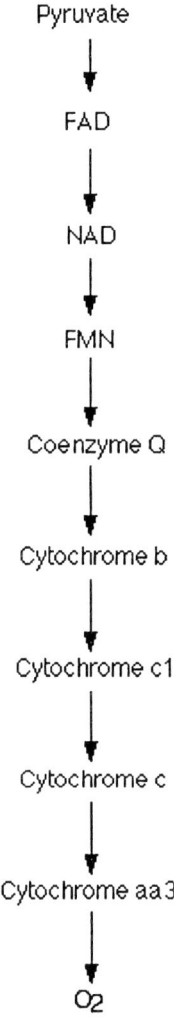

Figure 4.1. The steps in the respiratory chain in mitochondria.

The formation of superoxide is due to the auto-oxidation of some compounds such as coenzyme Q (ubiquinol), NADH dehydrogenase and cytochrome b. Numerous studies have been reported on the production of free radicals by mitochondria, but controversy remains because of variations in experimental conditions [24, 29, 56, 105]. The greatest amount of superoxide is produced at the end of phosphorylation and when succinate is the substrate. By adding blockers, such as rotenone or antimycin A, the formation of superoxide is further increased. But, cyanides, which inhibit cytochrome oxidase will totally block the formation of superoxide.

The amount of superoxide produced varies with the experimental conditions, but ranges from 0.20 to 1.0 nmoles of H_2O_2 per minute per mg protein. This represents 2% to 3% of the total oxygen consumption. These values also vary depending on the organ from which the mitochondria are derived [2]. Brain mitochondria produce more superoxide because it adds the activities of monoamine oxidase and dehydro-orotate dehydrogenase to the sources already mentioned. These are flavine enzymes that produce superoxide and hydrogen peroxide. The formation of hydrogen peroxide by different organs is presented in table 4.1. This variation is the result of different metabolic activity in different types of tissue. It is amazing to see the great amount of hydrogen peroxide formed during the physiological activities of these organelles [105, 205]. The studies of Chance, Oshino, and Loschen (Johnson Foundation and University of Tubingen) have explained the source and mechanism of hydrogen peroxide formation during physiologic activity of the mitochondrion. Indeed, the presence of superoxide dismutase (SOD) in the mitochondria has already been mentioned (chapter 3).

$$2O_2^{\bullet} + 2H^+ \text{—— SOD} \rightarrow H_2O_2 + O_2 \qquad (3)$$

Studies suggest the coupling of superoxide production with reaction (3) is a defense mechanism. The formation of superoxide is inevitable and its decomposition by SOD should benefit survival. However, hydrogenperoxide is also dangerous to the cell, so mitochondria also contain an enzyme to remove it. This in not catalase, as might be expected, but glutathione peroxidase (GSH Px), and enzyme that will decompose any kind of peroxide. Glutathione peroxidase catalyses the following reaction between glutathione (GSH) and hydrogen peroxide:

$$2\ GSH + H_2O_2 \text{—— GSH Px} \rightarrow GSSG + 2\ H_2O \qquad (4)$$

In this reaction, nature has chose the best way to get rid of the hydrogen peroxide. Compared with catalase, glutathione peroxidase has the advantage of being able to decompose very low concentrations of hydrogen peroxide. Secondly, glutathione peroxidase can decompose any kind of peroxide. Indeed, during mitochondrial respiration, a chemiluminescent emission and lipid peroxides are detected [29, 72, 78]. The formation of lipid peroxides during respiration is connected to structural chemiosmotic modifications of mitochondrial membranes.

The ability of mitochondria to swell and shrink is reversible up to certain limits. These conformational changes are strongly linked to calcium uptake, ATP biosynthesis, and inter-nucleotide transhydrogenation.

$$NADH + ATP + NADP \text{—— hydrogenase} \rightarrow NAD + ADP + NADPH$$

The intensity of calcium flux within mitochondria depends on the NAD(P)H/NADP ratio, which is also strongly related to the level of H_2O_2 [78, 103, 105].

Table 4.1 The *in vitro* formation of hydrogen peroxide by intracellular organelles. (after Cimino [34])

Organelle	Rate of H_2O_2 formation (nmoles/min per mg prot.)	Condition
Peroxisomes (rat liver)	0.9 - 1.3	Endogenous substrate
Peroxisomes (rat liver)	8.6 - 16.5	Uric acid
Microsomes (rat liver)	0.8 - 1.8	NADH
Mitochondria (pidgeon heart)	0.7 - 0.9	Succinate
Mitochondria (rat heart)	0.5	Succinate
Mitochondria (ox heart)	0.2	Succinate and antimycin
Mitochondria (rat brain)	1.2	Succinate and antimycin
Mitochondria (rat brain)	0.7	Succinate
Chloroplasts	400	Succinate and light

The formation of hydrogen peroxide and lipid peroxides during mitochondrial respiration has much greater consequences than expected. The release of hydrogen peroxide increases proportionally with the partial pressure of oxygen. Therefore, it is no wonder that during a cerebral or cardiac ischemic event, the concentration of hydrogen peroxide in the extramitochondrial space drops and, consequently, strongly influences oxidative phosphorylation and ATP levels [56, 123, 159]. Secondly, it was found that in old animals, a significant increase in lipid peroxide formation takes place in the mitochondria because of increased superoxide production: 2.54 nmoles/minute/mg protein compared to 1.9 nmoles/minute/mg protein [29, 103, 181].

Mitochondrial respiration and the formation of reactive oxygen species as a byproduct is an essential process for life. The formation of reactive oxygen species during respiration also regulates and influences all these essential processes. Therefore, it is no wonder that according to Professor Sohal the rate of mitochondrial formation of hydrogen peroxide is irreversibly related to life span [43, 180, 181]. We will discuss this more in chapter 9.

4.2 PEROXISOMES

In 1966, C. DeDuve reported the discovery of a new type of intracellular organelle in hepatocyte cytoplasm. They are called peroxisomes because they produce large amounts of hydrogen peroxide. The amount of hydrogen peroxide produced in peroxisomes is summarized in table 4.1. Peroxisomes have since been identified in kidney, gut, heart, lower organisms, and plants [47, 130, 144].

Peroxisomes contain a varied and rich range of specialized enzymes with particular roles in metabolism. The basic role of these enzymes is the formation of hydrogen peroxide as the consequence of oxidation of amino acids and α-hydroxy acids.

D(L) amino acids — oxidases → α-keto acids + H_2O_2
L-α-hydroxy acids — oxidases → α-keto acids + H_2O_2 (5)

Uric acid — uricase → alantoin + CO_2 + H_2O_2 (6)

Reaction 6 does not take place in man, as humans do not possess uricase. Due to the production of α-keto acids, the activity of peroxisomes is closely tied to the tricarboxylic cycle, which is located in mitochondria.

To remove the hydrogen peroxide produced in these reactions, peroxisomes possess an appreciable amount of catalase. This enzyme can catalyze a catalase specific reaction (7) or a peroxidase type reaction (8).

$2 H_2O_2 + O_2$ — catalase → $2 H_2O + 2 O_2$ (7)

H_2O_2 + phenols (ethanol) — catalase →
$2 H_2O$ + quinone (acetaldehyde) (8)

The second source of hydrogen peroxide production is the β-oxidation of fatty acids, an essential process that provides chemical energy and takes place in the mitochondria. It seems that peroxisomes act as a functional reserve in the organism. Following a diet rich in lipids, the amount of peroxisomes in the liver and the amount of fatty acid metabolized by these organelles increases significantly. For every 2 carbon residues from a fatty acid molecule, 1 molecule of hydrogen peroxide is formed. Therefore, following a diet rich in lipids, the formation of hydrogen peroxide might exceed the protective capacity of catalase, creating favorable conditions for the peroxidation of lipids from membranes, leading to cell lysis.

Dr. Queroga demonstrated the role of peroxisomes in the process of thermal regulation. Reactions (7) and (8) are exothermic, possessing a ΔH of -10 kcal and -20 kcal, respectively. Since the liver peroxisomes produce approximately 1.5 μmoles of hydrogen peroxide per minute per gram of liver, 7×10^{-2} cal/min per gram of liver will result. As liver has a water content of 69%, it follows that peroxisomal respiration produces a heat increase of approximately 0.1 cal/min/g of liver. This involvement of peroxisomes in thermoregulation is widely shared by organisms.

The bombardier beetle is able to eject a hot jet of peroxides as a form of defence. The bombardier stores separate solutions of 25% hydrogen peroxide and 10% hydroquinone. When in danger, the solutions are rapidly mixed in a chamber that is rich in peroxisomes and catalase. Reaction (8) takes place, resulting in a hot (80° C) stream that can be projected a great distance.

Peroxisomes are also involved in several metabolic processes having unknown consequences, such as:

- Biosynthesis of ether lipids and plasmalogens
- Oxidative degradation of dicarboxilic and monounsaturated acids
- Oxidative degradation of the acyl chain of xenobiotics such as aromatic fatty acids or hypolipemic drugs (clofibrate) [127]

All of these processes require the oxidative capacity provided by the release of hydrogen peroxide by peroxisomes.

4.3 PHAGOCYTOSIS

Phagocytosis, which takes place in polymorphonuclear leukocytes (PMNL) and tissue macrophages, is the best known metabolic process that provides free radicals (reactive oxygen species). Even though this process was discovered by Netchinikoff in the 19th century, it is only in the past 20 years that studies have begun to unravel the complex reactions that occur in all organisms [38, 125, 156].

Phagocytosis is the most important process in cellular immunity and consists of several steps. These are recognition of a pathogen, mobilization and transport of leukocytes towards the infection site by chemotactic signals, adherence to the microbial cell, endocytosis of the invader, and its destruction with hydrolytic enzymes. These steps are composed of several sequential biochemical reactions. The number of studies published about this process continues to grow.

Of the many and complex steps involved in phagocytosis, we will only discuss the role of reactive oxygen species, which is essential in the killing of pathogens. The following is a summary of the major steps.

The first step is *mobilization* of phagocytotic leukocytes (PMNL) that take place in response to a chemotactic signal. These "call for help" signals consist of small peptides like formyl-methionyl-leucyl-phenylalanine (fMLP). These compounds form a concentration gradient that can be detected at 10^{-8} molar.

The second step is *adherence* of leukocytes to the membrane of the pathogen. This is an extremely complex process involving various stimuli such as fMLP, phorbol esters, leukotriene B_4, Fc fragments, opsonins, components of complement (C5a), immune complexes, etc. The metabolic activation of leukocytes is a function of the number and affinity of the receptors on the leukocyte membrane. Therefore, we present in table 4.2 the best known chemoattractents and stimuli for phagocytosis. The great structural variety of chemoattractents and stimuli is apparent. Most chemoattractents possess a certain specificity for some cells, but there are also many nonspecialized stimuli. This is strong evidence that adherence and the triggering of leukocyte activation is a nonspecific membrane phenomenon. Supplementary proof is provided by the inhibition of phagocytosis by specific SH group blockers such a N-ethyl-maleimide.

Table 4.2 Chemoattractants and stimuli for phagocytotic leukocytes

Chemoattractants	Type of cell*
Compliment-derived factors; C567, C5a	N, M, E, B
Kinin precursors, kalikrein, fibrino-peptides	N, M B
Cell-derived factors (limphokin, IgM)	N, M
Neutrophile-derived factors	N, M, E, B, L, F
Denatured albumin and casein	N
Proteins from fibroblast culture	N, M
Lecithins	N
Peptides (fMPL), bacteria secreted factors	N, M, E, B
Bacterial lipids, phospholipids, lipopolysaccharides	N, M
Bioactive derivatives of arachidonic acid	N, M
Histamine and its metabolic products	E
Collagen and its degradation products	M, F

Soluble stimuli	Particulate stimuli
Concanavalin, phorboesters	Latex, polystyrene
Cytochanlasine	Zymosan (yeast)
Digitonin, Calcium inophore	Bacteria (Staphlyococcus)
Desoxycholate and electro-negatively charged detergents	Carageenan

*N=neutrophiles, M=monocytes, E=eosinophiles, B=basophiles, L=lymphocytes, F=fibroblasts

The third step is *transduction,* a natural consequence of the membrane effect triggered by the stimuli-receptor reaction on the leukocytic membrane. It triggers the sequence of reactions that lead to the activation of the leukocyte. As depicted in figure 4.2, as soon as the leukocyte adheres to the pathogen, NADPH oxidase and superoxide dismutase (SOD) are activated, producing superoxide ($O_2^{\cdot}$) and hydrogen peroxide that are released outside the cell. These reactive oxygen species act as a kind of chemical bullet to kill the pathogen through oxidative deterioration of its membrane (lipid peroxidation). The production of reactive oxygen species by the leukocyte continues while the pathogen is engulfed (pinocytosis) into the leukocytic cytoplasm as a phagosome.

All these steps require an extraordinary array of reactions all working in sequence. In order to keep this part of the discussion simple, we will mention only a few reactions. The activation of phospholipase A_2 and C releases arachidonic acid from membrane phospholipids. Arachidonic acid acts as the substrate for biosynthesis of prostaglandins and leukotrienes. The main source of reactive oxygen species is phosphorylated NADPH oxidase. The reactions that started in the membrane continue in the cytoplasm. When these activations and reactions occur, the release of chemical energy and the consumption of oxygen and glucose increases rapidly. Because of this process is called a respiratory burst. The following reactions are involved.

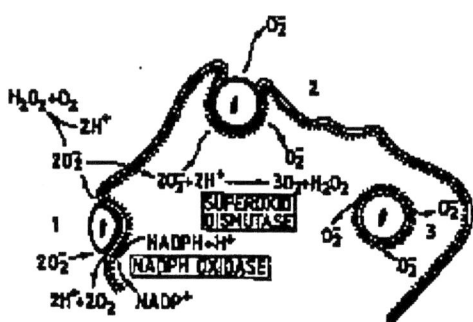

Figure 4.2. The major steps in phagocytosis. 1, adherence; 2, pinocytosis; 3, formation of a phagosome. Points at which the formation of reactive oxygen species may occur in phagocytosis are noted.

glucose + NADP — hexose monophosphate shunt → NADPH
O2 + NADPH — oxidase → NADP + $O_2^{\bullet}$ (superoxide)
$2\ O_2^{\bullet} + 2H^+$ — SOD → H_2O_2
$H_2O_2 + Cl^-$ — myeloperoxidase → OCl^- (hypochlorite) + $^1O_2$
$OCl^- + {}^1O_2$ + amino acids + fatty acids → chloramine + peroxides + aldehydes

These reactions provide energy and the killing agents necessary for structural modifications in phagocytosis. During this process a wide variety of reactive oxygen species are produced.

These reactions only summarize an impressive array of reactions that lead to the killing and destruction of pathogenic agents. This sequence of reactions, although it is programmed, is not rigid. Qualitative variations are possible based on the leukocyte and pathogen types. The presence of the above mentioned enzymes and their activation assures a rapid biological response as well as their regulation.

In addition to the killing systems based on reactive oxygen species, phagocytic leukocytes also use cationic proteins, lysozyme, histone, elastase, and hydrolytic enzymes for the destruction of the pathogen. The last step of the destruction of a pathogen by a leukocyte is its digestion by hydrolytic enzymes that takes place at an intracellular pH of 5.5, a pH rarely found in other cells.

Studies have demonstrated an impressive variety of compounds released by the activated leukocyte. The selection of compounds released by leukocytes and macrophages seems to depend on the type of cell, the nature of the pathogen, and the trigger or stimuli (table 4.2). This elasticity in response permits the phagocytizing cell to adapt to varied conditions. The response of the cell is not only to amplify the ability to destroy the specific pathogen, but the release of signals to attract more leukocytes (table 4.3).

This great variation in components released is valuable for the production of reactive oxygen species. Numerous experiments have demonstrated that the yield of reactive oxygen species varies. For superoxide it ranges from 10 - 100 nmoles $O_2^{\bullet}$/min per 10^7 cells. For hydrogen peroxide it ranges from 2 - 20 nmoles H_2O_2/min per 10^7 cells. In spite of the great variation in results due to the variety of analytical methods used, these data indicate an impressive amount of reactive oxygen species is released in the inflammatory process.

The release of reactive oxygen species during phagocytosis provides the best way to measure the response of leukocytes or tissue macrophages. Indeed the chemical determination of hydrogen peroxide or superoxide is not difficult, but it is time consuming. One popular method is chemiluminescence.

As shown in figure 4.3, as soon as a suspension of leukocytes or a diluted blood sample is activated by one of the stimulants listed in table 4.3, chemiluminescence begins, reaching a maximum after 10 to 15 minutes. It then slowly decreases. Chemiluminescent emission by activated leukocytes is due to reactive oxygen species and prostaglandins released during phagocytosis.

The activation of phagocytic cells represents a serious hazard to the integrity of the connective tissue matrix. The release of reactive oxygen species by PMNL can damage neighboring tissues that surround the infection site. The depolymerization of hyaluronic acid, proteoglycans, and collagen as well as the action of proteases are responsible for inflammation and destruction of several biological compounds. Inflammatory diseases will be discussed further in chapter 9.

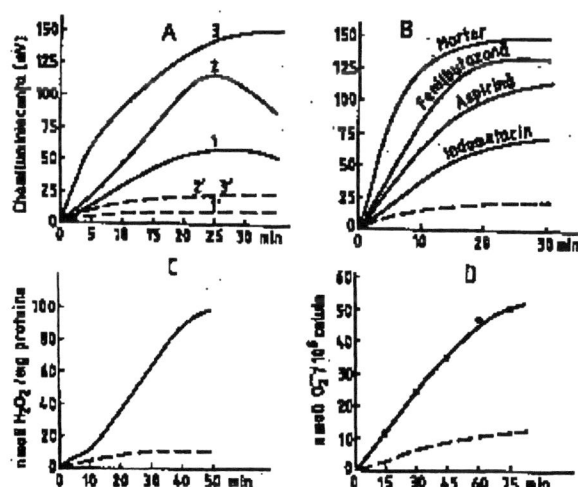

Figure 4.3. A, Chemiluminescence of stimulated (solid line) and nonstimulated (dashed line) polymorphonuclear leukocytes; B, Effect on chemiluminescence of antiinflammatory drugs: Control (Martor), Phenylbutasone (Fenilbutazona), Aspirin (Aspirina), Indomethacin (Indometacin). C, The release of hydrogen peroxide (nmoles per mg protein) in stimulated (solid line) and nonstimulated (dashed line) polymorphonuclear leukocytes. D, The release of superoxide (nmoles per 10^6 cells) in stimulated (solid line) and nonstimulated (dashed line) polymorphonuclear leukocytes. From a study published by Dr. Olinescu [123].

Table 4.3 Compounds released by activated phagocytic leukocytes

O_2 FREE RADICALS	ENZYMES

Superoxide	Lysozyme
Hydroxyl	Neutral proteases (elastase,
Hydrogen peroxide	collaginase, plaminogen)
Singlet oxygen	Acid proteases (phosphatase)
BIOACTIVE LIPIDS	COMPONENTS OF COMPLEMENT
Metabolites of arachidonic acid	C_1, C_2, C_4, C_{3b}, C_7
(prostaglandins, leukotrienes,	Factors P and D
thromboxanes)	Properdin
SLOW REACTING SUBSTANCE	FACTORS
PLATELET ACTIVATING FACTOR	Fibroblast stimulating factor
LIGAND PROTEINS	Endothelial cell stimulating factor
Transferrin	Tumor growth inhibitor
Fibronectin	Tumor necrotic factor
NUCLEOSIDES	Angiogenesis stimulating factor
Thymidin	HISTAMINE
Uracyl	
Uric acid	
Adenosine	

4.4 PROSTAGLANDINS AND LEUKOTRIENES

Arachidonic acid released from membranes is the main substrate for the biosynthesis of prostaglandins (PG) and leukotrienes (LK). This biosynthetic process takes place in the phagocytizing leukocytes, in the endothelial cells of blood vessels, seminal versicles, etc.

Von Euler began his studies of prostaglandins in the 1930s. However, it was only in the 1960s that the importance of these tissue hormones (autocoids) became clear. Today more than 10,000 paper and books deal with these eicosanoids (derivatives of arachidonic acid possessing a common basic structure).

Prostaglandins, leukotrienes, and thromboxanes (TX) are hormone like compounds called autocoids, acting in close proximity to the site of their synthesis. Arachidonic acid is present in the phospholipids of the membrane (phosphatidylcholine, phosphatidylserine) triglycerides, and cholesterol esters. Therefore, the first step in prostaglandin synthesis is the release of arachidonic acid from the membrane under the catalytic action of phosphlipase A_2 (PLA_2). Phospholipase is activated by several stimuli such as phagocytosis, antibodies, immune complexes, bacterial endotoxins, lymphokines, etc. While phospholipase A_2 (type I) is stimulated by calcium ions at physiologic pH values, phospholipase A_2 (type II), found in the cytoplasm, is inhibited by calcium and has a maximum activity at acidic pH.

Snake venom is rich in PLA_2. It acts by hydrolyzing phosphatidylcholine from the membrane, releasing lysolecithin. Lysolecithin has a high hemolytic capacity. Arachidonic acid is also released by thrombin from platelets. In the lungs it is released by releasing contracting factor and by slow reacting substance of anaphylaxis, especially in during bronchial asthma.

The involvement of free radicals in prostaglandin synthesis appears in the second step involving alterations in the structure of arachidonic acid. Two enzymes act on arachidonic acid to produce the two branches of reactions presented in figure 4.4. The existence of these two branches are important because each of them produces different products with opposite biological actions.

As shown in figure 4.4, the first, unstable, product ($PGGH_2$ or 5-HPETE) possess a peroxide (-OOH) group. Further branching can occur at this point. Unlike other biological processes in which free radicals are formed, prostaglandin synthesis produces only peroxides. The principle endoperoxide that is formed in this step is PGG_2, a very active biological compound. It is 10 times stronger than ADP in inducing platelet aggregation.

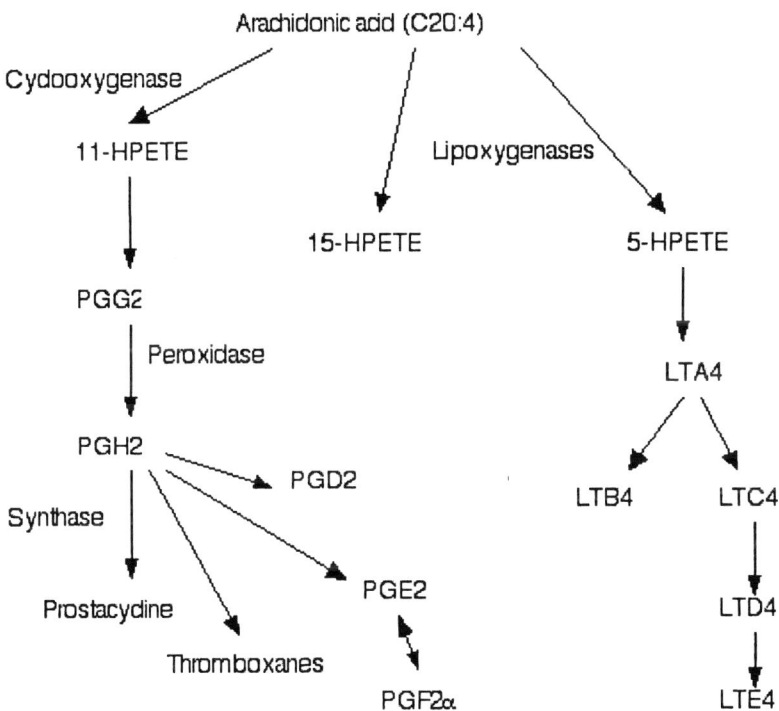

Figure 4.4. The arachidonic cascade showing the formation of prostaglandins, thromboxanes, and leukotrienes from arachidonic acid.

Another feature of eicosanoids that may be difficult to understand is the opposing properties of many of these compounds. For example, prostaglandin E_2 (PGE_2) has a different action in the inflammatory process. As shown in table 4.4, early in the inflammatory process, PGE_2 acts on cells within the vascular wall, acting as a hemostatic agent and is pro-inflammatory. Later, PGE_2 acts in an anti-inflammatory manner, decreasing leukocyte activation.

The involvement of reactive oxygen species in the biosynthesis of eicosanoids is due to the necessity for a rapid modification of arachidonic acid. This structural modification takes place during the catalytic action of cyclooxygenase, which also exhibits a peroxidase activity. In evidence of this it is noted that antioxidants completely inhibit cyclooxygenase.

One of the most important biological processes connected with reactive oxygen species and prostaglandins is platelet aggregation. Platelets are essential in the blood coagulation process. These sensitive cells are activated following reception of chemical signals released from activated macrophages, resulting in an influx of calcium ions. The activation of platelets takes place in their membranes and includes the inhibition of translocase and activation of phospholipase A_2. These modifications favor the action of exogenous ADP, which triggers increased receptivity towards other stimuli (adrenalin, collagen, thrombin, etc.) and releases ATP, ADP, serotonin, etc. from the platelets. This reaction cascade triggers the aggregation of platelets and their adherence to epithelial cells in the blood vessels. During this cascade, an important prostaglandin synthesis takes place with the cyclooxygenase branch reactions being predominate. The quantitative and qualitative regulation of this process is very important. Both endoperoxides PGG and PGE_2 have key roles favoring the shift of biosynthetic reaction chains towards thromboxane or prostacyclin (PGI_2). Prostacyclin and thromboxane have opposite effects, and the equilibrium between them constitutes the most important mechanism of hemostasis as it regulates arterial tonus and the adherence of platelets to blood vessel endothelial cells.

Table 4.4 Biological properties of prostaglandin E_2 (PGE_2)

Proinflammatory	Anti-inflammatory
Vasodilation	Vasoconstriction
Increase permeability of blood vessels	Bronchoconstriction
Activation of pain	Inhibition of phagocytosis
Chemotaxis	Inhibition of cytotoxicity
Stimulate leukocyte adherence	Stimulate cyclooxygenase activity
Stumulate pentose monophosphate shunt activity	Supress leukotriene formation (LTB_4)
Release lysosomal enzymes	
Increase C_{3b} receptor expression	

Platelets and leukocytes function together in the inflammatory process. Therefore, it is reasonable to expect a certain amount of communication between them would exist. This is the case as was shown by Marcus and Hajjer [109]. They showed that prostaglandin synthesis (the first and second steps, which include endoperoxide formation) takes place differently in these cells. Endoperoxide, PGH_2, and leukotriene A_4 act as chemical signals between leukocytes and platelets. Also, endothelial cells produce prostacyclin when endoperoxides from activated leukocytes cross their membrane.

Stimulated platelet-neutrophile suspensions can also generate lipoxins, biologically active eicosanoids derived from arachidonic acid by the 15-lipoxygenase pathway. In the neutrophile-platelet-endothelial cell interaction the donor is the neutrophile, which produces an epoxide intermediate (leukotriene). This epoxide in turn triggers the production of the leukotriene LTC_4, which has vasoconstrictive properties. This production is not inhibited by aspirin. In this process, red blood cells, once activated, produce LTB_4.

The reactions catalyzed by cyclooxygenase and prostaglandin synthase are specifically inhibited by antiinflammatory drugs (aspirin, indomethacin, phenylbutazone, and other nonsteroidal drugs). However cyclooxygenase's peroxidase activity is able to metabolize large range of compounds, including polycyclic hydrocarbons that are carcinogenic (benzopyrene). This activates the carcinogenic activity of these compounds because it is the products that strongly bind DNA [31, 93, 108].

Prostaglandin biosynthesis has an important role in the organism. In vitro studies estimate the magnitude of prostaglandin synthesis in platelets as 1 to 40 ng of PGE_2 per 5×10^7 platelets. This process may be occurring simultaneously in several places in the body [108, 141]. Intermediates in the metabolism of arachidonic acid are endoperoxides that, in spite of their short life, are strongly involved in the regulation of eicosanoid biosynthesis.

4.5 PEROXIDATION OF UNSATURATED LIPIDS

The peroxidation of unsaturated fatty acids is the best studied of the natural processes involving the formation of free radicals. As discussed in chapter 3.6, the peroxidation of organic molecules is the consequence of reactive oxygen species formation. The peroxidation of structural components is likely to occur when the antioxidant systems are overwhelmed. Peroxidation is a universal phenomenon that occurs in living organisms as well as in manufacturing processes (of rubber, plastic, drugs, etc.).

Here we examine the question of peroxidation as a natural process, physiologic process. This is, in fact, a controversial question, but the majority of published studies seem to suggest the answer is yes. One problem that has not been resolved is the magnitude of this process. In part, this is because the answer depends on the techniques used to measure the formation of free radicals [9, 36, 45, 75, 80, 123, 129, 203].

As presented in figure 4.5, the entire process of reactive oxygen species formation, including peroxidation and peroxide decomposition, can be detected through a large array of techniques (physical, chemical, radioimmuno assay, chemiluminescence, enzymatic, and chromatographic methods). The sensitivity of these techniques, the availability of good instruments, and experimental conditions are all factors that strongly influence the final results. This influence is explained by the characteristics of reactive oxygen specie components: dynamic formation, very short half-life, rapid intertransformation, and presence of antioxidants or other interfering substances. When these measurements are to be made in a biological system, the problems are magnified.

A method that has been used in thousands of studies is the thiobarbituric acid reaction. Thiobarbituric acid (TBA) reacts with malondialdehyde (MDA), a decomposition product of lipid peroxides. TBA also reacts with other components from plasma, such as some carbohydrates, forming products that absorb at other wavelengths, but that may interfere with the test. In spite of these problems, this method is still widely used. The result is given in an amount of thiobarbituric acid reactive substances (TBARS). This method is still in use because other, more accurate, methods for the measurement of peroxidation, such as the measurement of ethane or propane in exhaled air is both expensive and the instruments are not widely available. This lack of specificity is a problem shared by most of the techniques listed in figure 4.5. Each method presented in figure 4.5 measures an intermediate from the chain of reactions involved in free radical formation or degradation. The lack of specificity when making measurements in biological systems is likely to be a problem that cannot be avoided, since all reactive oxygen species are capable of reacting with compounds in biological fluids.

In spite of problems with the techniques in regard to detection limits, all of these techniques have demonstrated the presence of reactive oxygen species in biological systems, including human systems. To illustrate the problems inherent in this research, table 4.5 lists the normal lipid peroxide values in humans. From this table it is apparent that the measurement of peroxide and peroxide-like compounds is dependent on the sensitivity of the method. Because of the expense and time requirements of high-level detection techniques, the large numbers of samples necessary for quality statistical analysis often cannot be done. Instead, most techniques depend on matched controls to detect significant changes. In spite of these problems, both old and new techniques confirm that peroxide-like compounds exist in blood plasma and urine. The low level of these compounds shows the efficiency of the protective antioxidant systems.

Table 4.5 Normal values for lipid peroxides and its derivatives in human plasma

Compound	Amount (μM/L)	Method*	Author
TBARS*	1.1	TBA	Krinsky (92)
	3.2	TBA	Biaglow (20)
	3.1	TBA	Tara (187), Yagi (203)
	1.3	TBA	Duthie (55)
Peroxides	0.5	Cyclooxygnease	Miquel (114)
Peroxides	0.3	HPLC and CL	Frei (62)
4-hydroxynonenal	0.65	HPLC from urine	Yu (206)
Hydrogen peroxide	100	RIA	Varma (193)
Thymidin glycole	100	HPLC	Ames (5)

*Abbreviations: CL=chemiluminescence, HPLC=high performance liquid chromatography, RIA=radioimmunoassay, TBA=thiobarbituric acid reaction, TBARS= thiobarbituric acid reactive substances (generally expressed as malondialdehyde or MDA)

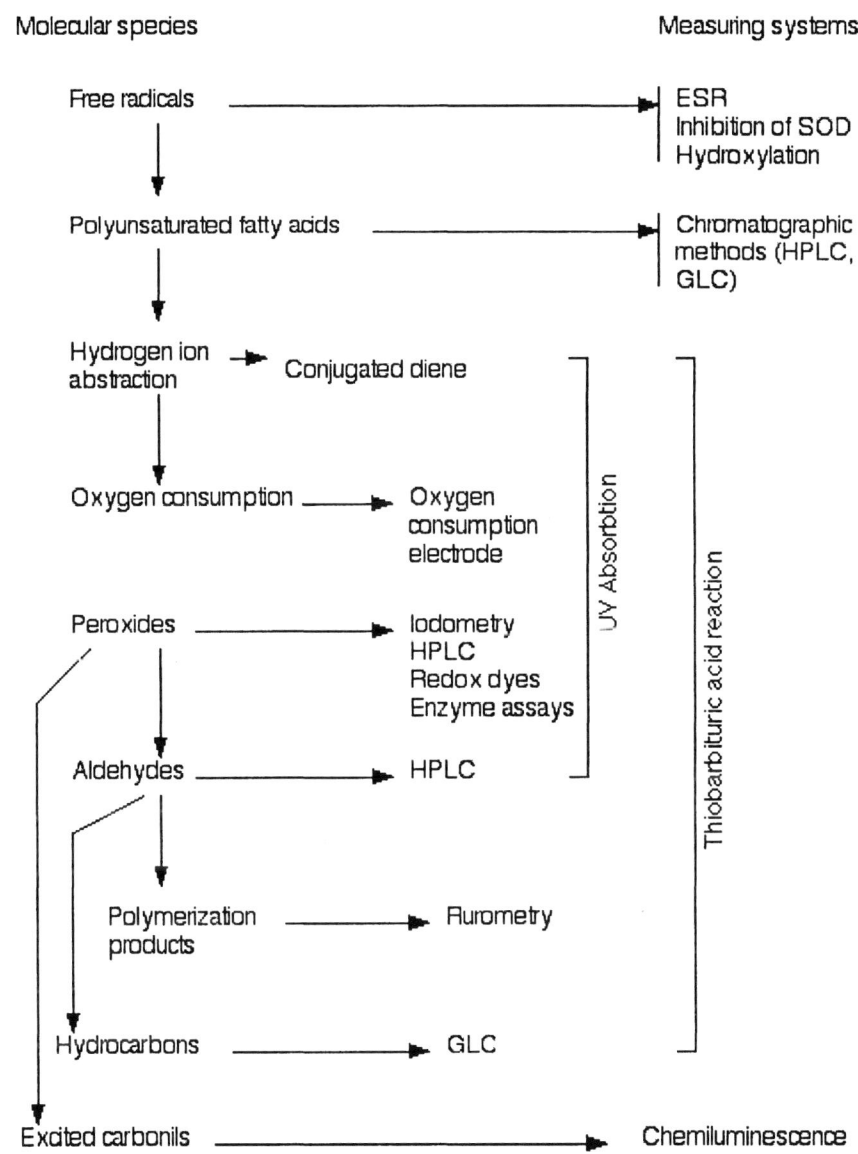

Figure 4.5. The evolution of reactive oxygen substances and applicable measurement techniques. ESR = electron spin resonance; GLC = gas-liquid chromatography; HPLC = high pressure liquid chromatography; SOD = superoxide dismutase.

As discussed in previous chapters, polyunsaturated fatty acids (PUFA), such as linoleic and arachidonic acids, are the most likely targets for the formation of reactive oxygen species. Consequently, cellular membranes, which are rich in PUFA, are the major site of peroxidation and most peroxide-like products are lipids (Fig. 3.2). Therefore, a correlation should exist between the low levels of lipid peroxides and the plasma concentration of PUFA. Indeed, the physiologic level for esterified PUFA is about 13 mM, while the nonesterified fatty acid level is about 0.5 mM and the phosphatdylcholine level is about 0.23 µM (9, 80, 123). A direct relationship was not found because of different homeostatic conditions.

Many reported studies that were performed using the TBA method showed that the low level of TBARS is increased 30% to 50% under heavy physical exertion [19, 55, 124] and during pregnancy, especially if toxicity is occurring [39, 89, 123, 203]. In these physiologic states, an important number of cells are destroyed and structural modification of tissue takes place, resulting in membrane disruption and the release of PUFA. Therefore, the evidence indicates that, under physiologic conditions, the sources of free radicals are strongly connected with the activation of oxygen as described in chapters 3 and 4.

- Mitochondrial respiration provides about 2 nmoles H_2O_2/ minute per mg protein [105]
- Peroxisomes release approximately 1.5 µmoles H_2O_2/ minute per mg protein [130]
- Prostaglandin biosynthesis produces between 1 and 40 ng PGE/10^7 cells
- Phagocytosis by leukocytes and macrophages release 10 to 100 nmoles superoxide per minute per 107 cells and 2 to 20 nmoles H_2O_2/ minute per 10^7 cells as well as important amounts of peroxides [89, 125]
- Peroxidative destruction of proteins and amino acids produce about 3 nmoles TBARS/ minute per mg protein [114, 150, 178, 180, 203]

In spite of the fact that most of this data were obtained with *in vitro* experiments, the magnitude of reactive oxygen species under physiological conditions is impressive, especially when extrapolated to the whole body scale. Also consider that under extreme, but still physiologic, conditions, such as heavy exertion, strong emotional stress, or pregnancy, the production of reactive oxygen species increases significantly due to the release of PUFA from cell membranes.

As Slater [177, 178] emphasized at the beginning of the 1980s, scientists realized that the decomposition products of peroxides are dangerous to the organism as they can inhibit several enzymes, especially those that require cell membranes to function. This may suggest a role for the presence of increased amounts of peroxides (as measured by TBARS) in many diseases or pathological conditions.

Physiological reactions may be strongly influenced by environmental factors [15]. Several reports emphasize the harmful effects of estrogens and estrogen-like compounds on organisms, beginning with wild aquatic forms and ending with humans. Many natural estrogens (estradiol, estrone) or synthetic estrogens (diethyl stilbesterol, 17-ethynylestradiol) are widely used as contraceptives on anabolic hormones. But many natural or synthetic compounds are present in the environment that can mimic estrogen in the body and can form quinones: DDT (tricholroethane), polychlorinated biphenyls, insecticides (atrazine, endosulphan), organic chlorinated compounds, polycarbonate plastics, and aromatic compounds (dibenzoanthracene, benzopyrene). As discussed in chapter 2, semiquinones are produced form aromatic compounds during metabolism by microsomal cytochrome P_{450}. Another source of free radical semiquinones are the enzymes leukocyte peroxidase and prostaglandin synthase. Estrogens are metabolized in a redox cycle while releasing hydrogen peroxide as a byproduct. This peroxide stimulates prostaglandin biosynthesis in the uterus and may be involved in the etiology of some uterine and breast cancers. In order to act, these chemical pollutants first bind to the estrogen receptor, triggering their metabolism [116, 127, 180, 207].

Both leukocyte peroxidase and prostaglandin synthase are able to use estrogen-like compounds as substrates, producing peroxides. The harmful consequences of estrogen pollution include male sterility in many organisms, including humans.

Fortunately, it has been discovered that a wide range of natural compounds called phytoestrogens also bind to the estrogen receptor. Phytoestrogens possess a significant antioxidant activity and thus do not form free radical semiquinones. Such beneficial phytoestrogens exist in important amounts in soy beans, licorice, ginseng, parsley, carrots, thyme, wild yam, and medicinal herbs such as chaparral and sasparilla. Japanese women in menopause do not have hot flashes and osteoporosis. This effect has been explained by their diet high in phytoestrogens from tofu, miso, and shoyu. Japanese women also have the lowest rate of breast and uterine cancer [116, 127].

4.6 ARE OXYGEN FREE RADICALS CHEMICAL MESSENGERS?

Our knowledge of free radicals is dependent on scientific research. Until the 1960s, most scientists did not accept the existence of free radicals. In the 1980s, the existence of free radicals was accepted to occur only under certain physiologic conditions. Only in the mid 1990s was evidence found that supports their existence under all physiological conditions. During this time it has been demonstrated that some free radicals, mostly peroxides and aldehydes, are able to cross the membrane and to travel distances as great as 60 to 3000 nm in the intracellular space or cytoplasm. Therefore, it is time to reconsider some knowns about natural processes. Some suggested revisions are:

- The transformation of epidermal cells increases linearly with the amount of superoxide produced extracellularly and superoxide dismutase has an inhibitory, regulative role [2, 150].
- Most DNA mutations are closely related to the metabolic processes that produce superoxide, such as mitochondrial respiration [56, 58, 181]
- Following exposure of epidermal cells or leukocyte cells to phorbol myristate (carcinogenic), a chemiluminesent emission takes place. This emission is inhibited by superoxide dismutase, retinoic acid, and antiinflammatory drugs [89, 156].
- Superoxide radical is produced not only by PMNL and macrophages, but by muscle cells, skin fibroblasts, endothelial cells, and B lymphocytes. During phagocytosis, superoxide is strongly involved in the biosynthesis of chemotactic factors from arachidonic acid [2, 67, 155]
- Superoxide released from macrophages or endothelial cells modifies LDL and the oxidized LDL is phagocytized by macrophages leading to the formation of arterial plaques [49]

- Superoxide participates in platelet aggregation and their adhesion to blood vessels
- Superoxide is also involved in the biosynthesis of NO (EDRF). So, generation of superoxide is involved in the regulation of the contraction of blood vessels [145]

According to the above evidence, it seems that superoxide should be included as a second level chemical messenger [14]. Many of the reactive oxygen species can function as intracellular messengers for chemoattraction, but there is evidence that reactive oxygen species might also function as second messengers involved in mitogenic stimulation of cultured cells. No only do reactive oxygen species lead to mitosis, but mitogens stimulate the production of reactive oxygen species [28, 158].

4.7 APOPTOSIS

The latest and very exciting cellular field in which reactive oxygen species are involved is cell death. This unavoidable natural process can occur by either apoptosis or necrosis. Necrotic death occurs by the breakdown of intracellular structures and lysis. Apoptosis is an active process of self destruction. Apoptotic death is characterized by shrinkage, membrane blebbing, and nuclear DNA fragmentation. Apoptosis occurs under a variety of physiologic conditions, including embryogenesis, metamorphosis, and cytotoxic T-cell mediated killing. It can be induced in normal or malignant cells of lymphatic origin by ionizing radiation or glucocorticoids.

Forrest et al. [61] of the Naval Medical Research Institute (Bethesda) demonstrated that hydrogen peroxide of an oxidative stress condition are able to induce apoptosis. Electron microscopic studies revealed morphologic changes characteristic of apoptosis in mouse thymocytes treated with hydrogen peroxide. A water-soluble derivative of vitamin E (Trolox) can inhibit and prevent hydrogen peroxide induced apoptosis.

As oxidative stress might be reached through various physiologic and pathologic conditions (see chapter 5), and important role for reactive oxygen species should be accepted. Any condition that generates a strong, continuous oxidative stress should accelerate apoptosis. Indeed, new studies have shown an accelerated cell death resembling apoptosis in extreme conditions such a viral or hepatocarcinogenesis or in pathologic neurodegeneration related to neurotoxins [28, 150].

Chapter 5

HARMFUL EFFECTS OF FREE RADICALS

Due to their high reactivity, free radicals are blamed for many harmful events within living organisms. The involvement of free radicals in the damaging actions resulting from exposure to ionizing radiation, as was discovered in the 1960s, has triggered extensive studies concerning the deleterious effects of free radicals at the molecular level. These studies have examined the effect of free radicals on proteins, nucleic acids, and especially on sulphur containing compounds such as cysteine and glutathione. In the 1970s it was demonstrated that cellular and sub-cellular membranes are the principle target of free radicals leading to harmful biological events within the organism.

5.1 OXIDATIVE STRESS

The concept of oxidative stress developed in the mid-1980s due to the work of numerous scientist, such as H. Sies (Germany), T. F. Slater (England), S. Orrenius (Sweden), and R. Sahal (USA). The differences between their definitions mostly relate to the extent of this process.

According to Sies [167], oxidative stress includes all oxidative damage produced by reactive oxygen species, including damage to proteins, lipids, and nucleic acids. This damage is sufficient to overcome the organisms antioxidant defenses. This restricted definition implies that oxidative stress occurs at the cellular level and is expressed as inflammation, carcinogenesis, cellular lysis, modification of energy producing processes, etc.

However, oxidative stress may also appear at the level of the entire body. This extended oxidative stress is the result of oxidative stress occurring at the cellular level. This complicates the definition of oxidative stress. In fact, oxidative stress may not be simple to appreciate as it possesses a dynamic character in which the antioxidant systems of the entire body are mobilized to defend against the threat. Therefore, in the early (reversible) period of oxidative stress, a diet well supplemented with antioxidants should be of help to the defense of the organism.

As will be discussed in chapter 8, organisms must cope daily with oxidative stress. If oxidative stress prevails, there may be pathologic results, as discussed in chapter 9.

However, the antioxidant systems of the body are normally sufficient to prevent serious damage from occurring. In an attempt to better understand these interactions, we will now present some oxidative stress reactions occurring at the cellular level.

5.1.1 Erythrocytes and Free Radicals

It should be no surprise that erythrocytes have been extensively studied. These cells are, in fact, nearly ideal to study as they have no nucleus, mitochondria, or other subcellular structures, and have a life span of 120 days. In addition to these advantages, these cells can be easily obtained in homogenous suspension. Thus, simple experimental models an be obtained that are useful for studying the action of drugs, xenobiotics, and the regulation of some metabolic processes.

A range of drugs, chemical pollutants, and toxic substances act on the erythrocyte membrane or on hemoglobin producing intermediate free radicals. These lead to the formation of methemoglobin, structural modifications, and hemolysis. This action may be rapid and total (as with phenylhydrazine or nitrates, or partial, often occurring step wise, as a function of concentration. In these later events it is possible to observe the defensive reactions of the erythrocyte come into play [84, 123].

A typical example is the action of menadione or vitamin K_3 (2-methyl-1,4-naphthoquinone) on an *in vitro* suspension of erythrocytes. At high concentrations (10^{-3} M), erythrocytic glycolysis is inhibited; ATP cannot be synthesized and glutathione and other sulphur containing compounds are oxidized. Consequently, methemoglobin formation and hemolysis will occur. But, at lower concentrations of vitamin K3 (10^{-6} to 10^{-4} M), the erythrocytes defend themselves from this attack by activating the hexose monophosphate shunt and increasing glucose consumption, ATP synthesis, and glutathione reduction. The end result is to reduce methemoglobin production, preserving the integrity of the membrane. Some partial reactions are (Hb = hemoglobin, met-Hb = methemoglobin, vK = vitamin K_3, glu = glucose):

$$Hb + vK \rightarrow met\text{-}Hb + vK \text{ epoxide} \tag{1}$$

$$met\text{-}Hb \xrightarrow{\text{reductase}} Hb \tag{2}$$

$$glu \xrightarrow{\text{hexose monophosphate shunt}} lactate + NADH + NADPH \tag{3}$$

$$\text{oxidized glutathione} \xrightarrow{\text{GSH reductase}} 2 \text{ reduced glutathiones} \tag{4}$$

In this example of oxidative stress, methemoglobin must be reduced so that the cell can survive. This is only possible for low concentrations of vitamin K3 as higher concentrations inhibit the hexose monophosphate shunt (reaction 3), which is needed to provide energy to the system. A clinical application of this is toxic methemoglobinemia. Treatment for this condition is venous injection of the dye methylene blue. This dye stimulates the shunt (reaction 3), providing the NADPH needed to drive reactions 1 and 2 [55, 123].

An important step in understanding the involvement of oxidative stress in erythrocyte metabolism involved the use of ESR, HPLC, and other instruments [17, 175]. This allowed reaction pathways to be followed after administration of aromatic compounds (aniline, nitrosobenzene). The result was the accumulation of organic peroxides, endoperoxides (15-HEPETE), and sulphydryl groups from hemoglobin [84]. The toxicity of these aromatic compounds manifests itself at different concentrations and in different organs. Toxicity resulting in anemia is due to their metabolism in the liver, resulting in compounds that later destroy the erythrocytes [17, 84, 127]. Exposure to high amounts of aromatic compounds may produce the following reactions:

- Formation of nitrosobenzene and nitrobenzene in the liver
- The transport of these metabolites in the blood
- Hemolysis due to:
    - peroxidation of lipids in the membrane
    - increased influx of calcium into the cell
    - decreased intracellular glutathione
    - compensatory increase of the hexose monophosphate shunt activity
    - formation of methemoglobin
    - formation of Heinz bodies and hemochrome
- hemolysis

The key reaction related to the survival of erythrocytes under oxidative stress is the decrease of glutathione (GHS). The cell is able to defend itself by regenerating glutathione. Once the formation of Heinz bodies has begun, the damage due to hemolysis is irreversible.

Hemoglobin is damaged by aromatic compounds at the position of the 93- and 104- cysteine residues. This triggers the precipitation of modified hemoglobin. Some aromatic metabolites possess a free radical characteristic. This increases the potency of the compound. People most susceptible to the damaging effects of exposure to aromatic compounds have a congenital deficiency in glucose-6-phosphate dehydrogenase or glutathione reductase. Both of these deficiencies hamper the antioxidant defense of the cell. People with thalassemia or sickle cell anemia possess erythrocytes with greater sensitivity to oxidants. Peroxidative reactions are involved in a number of red cell disorders, many of whom are inherited (table 5.1).

It is well known that resistance to malaria is selective against sickle cell disease. The benefit is this situation seems to be that the presence of the malarial organism in the erythrocyte induces oxidative stress systems, protecting the cells from the effects of reactive oxygen species [84, 123]. Antimalarial drugs (primaquine, pamaquine) when metabolized and after penetrating the erythrocyte produce a mild oxygen stress condition that kills the parasite but not the erythrocyte, because of its abundance of antioxidant enzymes.

Oxygen stress also occurs in new-born bilirubinemia. Before a child is born, he is adapted to low oxygen partial pressure. The higher oxygen partial pressure he is exposed to after birth causes oxygen stress. Since the newborn's antioxidant enzymes are also low, hemolysis occurs. Hemoglobin is then degraded to bilirubin, further metabolized and excreted. In some newborns, the bilirubin level can become quite high.

Table 5.1 Congenital and aquired disorders of human erythrocytes

Disease	Factor
Steatorrhea	Vitamin E
Acanthocytosis	
Sickle cell anemia	
Hemolytic anemia (some cases)	Glutathione
Oxoprolinuria	
Nonmalignant deficiencies	
Glucose-6-phosphate dehydrogenase deficiency	Decreased NADPH production
Acatalasemia	Catalase
Sickle cell anemia	Hemoglobin abnormality
Thalassemia	Reduction of hemoglobin synthesis
Paroxysmal nocturnal hemoglobinuria	Membrane defect
Erythropoetic protoporphyria	Porphyrin metabolism
Drug-induced hemolysis (espeically in patients with steatorrhea or G-6-phosphate dehydrogenase deficiency	Antimalarials Antipyretics Analgesics (aspirin) Sulfones (daspsone) Phenylhydrogene

Erythrocytes possess a high amount of glutathione and antioxidant enzymes (catalase, glutathione peroxidase) that allow the cells to decompose reactive oxygen species produced during oxygen transport (fig. 5.1). This explains their long life (120 days) while constantly exposed to an oxidizing environment, their resistance to mild oxidative stress conditions.

Reactive oxygen species are involved in the aging and death of erythrocytes. As erythrocytes approach the age of 100 days, the glutathione and antioxidant enzyme content decrease. Therefore, these aging erythrocytes are less resistant to oxidative stress and more susceptible to lysis and proteolysis. Microscopic examination finds a significant accumulation of Heinz bodies and hemochrome within the cells.

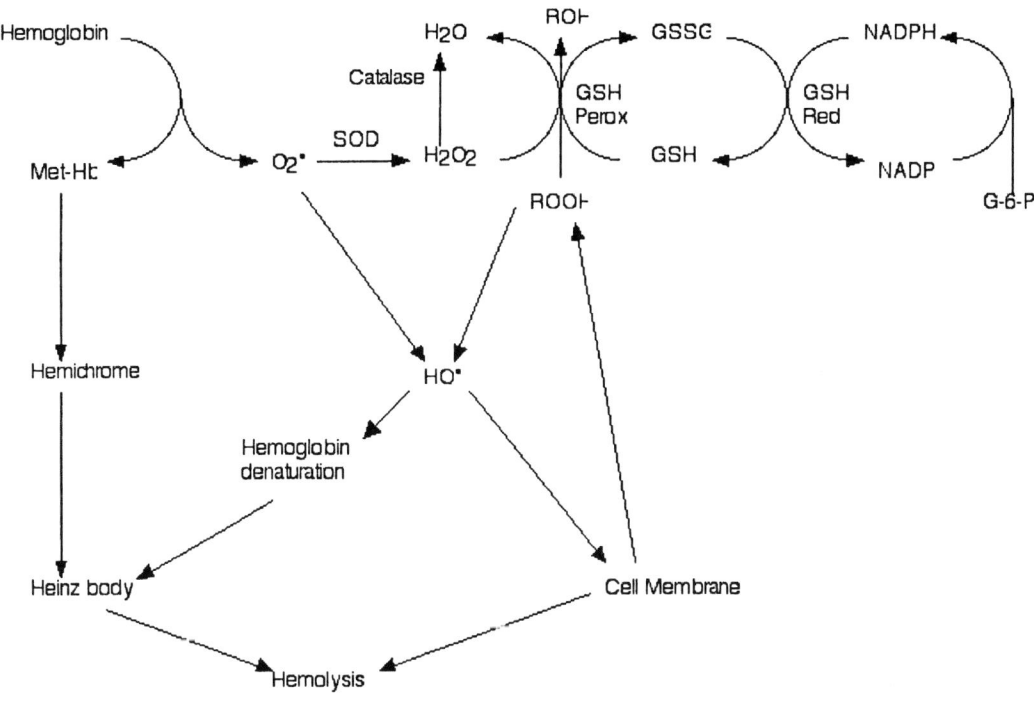

Figure 5.1. The enzymatic antioxidant systems in the erythrocyte and the role of hemoglobin in cell lysis. SOD = superoxide dismutase; GSH = glutathione; GSSG = oxidized glutathione; GSH Perox = glutathione peroxidase; GSH Red = glutathione reductase.

5.1.2 Liver and Decreased Glutathione

The liver plays a major role in helping the organism defend itself against oxidative stress. This function is mainly based on the liver's ability to metabolize a large array of endogenous and exogenous compounds. Studies confirm that the liver is continuously exposed to oxidative stress as it metabolizes aromatic compounds, carcinogens, drugs, etc. with the resulting formation of free radicals [5, 42, 178]. As described in a previous chapter, aromatic compounds produced as intermediate semiquinones are metabolized by NADPH dependent cytochrome P_{450} and flavoproteins. These reactions are not completely predictable, leading to the formation of semiquinones, autooxidative metabolites, and cross or branch reactions [39, 82, 127, 207].

Many aromatic compounds that are metabolized into semiquinone free radicals possess an electrophilic character and so will preferentially react with nucleophilic substances in cells. Nucleic acids and glutathione are preferred. Reduced glutathione (GSH), by reacting with these free radicals, protects other, more important, compounds within the cell: nucleic acids, sulphur-containing enzymes, and lipid membranes. These reactions have been demonstrated for endogenous (such as vitamin K_3) and exogenous (such as aniline) aromatic compounds and organic solvents (such as chloroform or toluene).

$$2 \text{ aromatic} + 2 \text{ GSH} \rightarrow 2 \text{ aromatic-SG} \tag{5}$$

$$2 \text{ aromatic-SG} \rightarrow 2 \text{ aromatic-OH} + 2 \text{ GSH} + H_2O_2 \tag{6}$$

Reaction (6) also produces 4 reduced flavin proteins. The aromatic compound is metabolized into a non-toxic, hydroxylated compound (aromatic-OH) that is more water soluble than the original compound and thus more easily excreted. The hydrogen peroxide produce in the reaction is decomposed by catalase.

These aromatic compounds and organic solvents are mainly metabolized in the liver, and a drop in hepatic glutathione has been demonstrated during acute exposure to these compounds [6, 55, 127, 178]. In one experiment, the liver from a live rat was perfused with an aromatic peroxide and the resulting chemiluminescence was recorded. Chemiluminescence is produced when semiquinone free radicals (Q$^\cdot$) react with oxygen, producing superoxide [30]. The presence of superoxide dismutase (SOD) directs the final reaction toward non-toxic reduced quinones [186].

$$Q^\cdot + O_2^{\cdot -} + 2H^+ \xrightarrow{SOD} QH_2 + O_2$$

Many experiments with toxic aromatic compounds, organic solvents, and chemical pollutants have demonstrated a quantitative relationship between the formation of aromatic free radicals, the decrease in hepatic glutathione, and the extent of damage to hepatic parenchyma [6, 12, 90, 127].

As seen in table 5.2, after acute intoxication with allyl alcohol, significant liver damage occurred as shown by the release of glutamate-pyruvate transaminase (GPT) and the formation of thiobarbituric reactive substances (TBARS) in direct proportion to the decrease in glutathione and the antioxidant vitamins C and E.

5.2 CELL MEMBRANE DAMAGE

Free sulphydryl groups of low molecular weight thiols and sulphydryl proteins are the main target of reactive oxygen substances. At the cellular level the preferred target groups are free sulphydryl groups located within the cellular membrane. The cell membrane is a bilayer of phospholipid with the fatty acids pointing inward. This forms a hydrophobic region in the center of the membrane while the hydrophilic portion of the molecules face outward in both the inner and outer sides of the membrane. Proteins are embedded in the lipid bilayer, some of which extend completely through the membrane. Many of these proteins act as regulated pores, allowing the passage of ions, water, etc.

Pore proteins are opened and closed by the action of stimuli such as biological messengers. These proteins have a significant content of sulphydryl groups. The lipid bilayer is a flexible structure, that can be penetrated by free radicals or other reactive oxygen species, allowing the attack of sulphydryl groups on proteins embedded in the membrane. Vitamin E is a lipid soluble antioxidant that is believed to be very important to protect the membrane from the action of reactive oxygen species. The loss of cell viability only occurs after the loss of a certain amount of vitamin E by the cell (to the attack of free radicals). It has been specifically shown that vitamin E protects the calcium-dependent ATPase by maintaining the sulphydryl groups of the protein in a reduced state.

Table 5.2 Variation of some biochemical parameters from rat liver following toxic exposure to allilic alcohol (1.5 mmol/kg) and the protective role of vitamin E [35].

	Vit. E deficient		Vit. E supplemented		
	Time (hours)		Time (hours)		
Treatment	0	1	0	1	2
Vit. E (pmol/mg prot.)	4.2±0.7	0.5±0.1	307±36	322±50	356±69
GSH[a] (nmol/mg prot.)	26.7±3.1	2.8±0.2*	23.9±2.6	6.6±0.9*	6.4±1.2*
TBARS[a] (pmol/mg prot.)	8.3±0.4	500±54*	7.2±0.5	16±9	5±1
Vit. C (nmol/mg prot.)	4.3±0.4	3.5±0.2*	6.0±1.2	3.9±0.3*	3.5±1.2*
Serum GPT[a] (units/ml)	40±12	1322±402*	31±4	67±19	284±65*

[a] GSH = Glutathione, TBARS = Thiobarbituric acid reacting substances, GPT = Glutamate-pyruvate transaminase
*Significant difference compared to 0 time, $p<0.05$

5.2.1 Increased Capillary Permeability

Inflammatory processes lead to an increase in vascular permeability, arterial constriction, and macromolecular penetration of the vascular wall. Therefore, it should be expected that the presence of reactive oxygen substances should influence capillary permeability. This has been well demonstrated by Parks and Granger [132, 133]. They introduced a superoxide-generating system consisting of xanthine and xanthine oxidase into the facial pouch of a hamster. As soon as 2 minutes after initiation of the experiment, labeled dextran began to permeate the vessels. The rate maximum was reached in 10 minutes. Additional evidence was obtained by the addition of superoxide dismutase, catalase, or hydroxyl radical scavengers such as methionine, mannitol, or dimethylsulfoxide, which inhibited the permeation of the dextrose.

In their next experiment, Parks and coworkers injected dogs with 5 mg α-naphthyl thiourea (ANTU), which produces pulmonary edema. They then measured parameters related to vascular permeability such as lymph flow rate (J), ratio of protein concentration in lymph (C_L) and plasma (C_P), and the amount of extravascular water in the lungs (Q_w) (see table 5.3). When the left arterial pressure is surgically increased, the C_L/C_P ratio decreases to 0.4 while the lymph flow increases four fold. ANTU administration increases lymph flow rate nine fold, while the increased C_L/C_P shows the increase in protein permeability. The most effective antioxidants were superoxide dismutase and dimethylsulfoxide, which prevent the involvement of superoxide radical. This supports the observation that where there is limited ischemia with reversible lesions, the mobilization of leukocytes takes place with a significant release of superoxide dismutase.

Table 5.3 Modification of vascular permiability in lung following tha administration of ANTU and antioxidants [132].

Conditions	J (µl/minute)	C_L/C_P	Qw (g/g)	Clearance (µl/minute)
Control (resting state)	14	0.7	3.9	10.6
IAP*	56	0.4	4.7	15.8
ANTU + IAP	120	0.6	7.0	97.1
ANTU + IAP + DMSO	225	0.5	6.0	75.4
ANTU + IAP + SOD	180	0.5	4.6	62.8
ANTU + IAP + Catalase	225	0.7	8.0	48.5
ANTU + IAP + Leukocyte suspension	94	----	4.8	----

*Increased arterial pressure

The action of animal or reptile venoms is less well studied regarding the involvement of reactive oxygen species. Snake venoms produce a variety of events that are near instantaneous or slow acting. These effects are mediated through polypeptides having a molecular weight of less than 10,000 and slow reacting substances that are derived from unsaturated fatty acids. The increase in vascular permeability is very quick and results in the activation of phospholipase A_2, leading to the release of lytic factors. Coagulating factors and kinins are also released. Prevention of the hypotensive effect of snake venoms is possible through the use of nonsteroidal antiinflammatory drugs (phenyl butazone and indomethacin), which are also antioxidants.

5.2.2 Cell Lysis

Damage to cellular membranes by reactive oxygen species is controllable up to a threshold when a cascade of events occur. Increased membrane permeability allows the increase in intracellular calcium and the activation of proteases and phospholipases. Increased cytosolic calcium appears to be the trigger for massive cell and tissue damage that can occur with or without the involvement of reactive oxygen species [39, 72, 167, 190].

A correlation has long been suspected to be present between cellular necrosis and calcification. The free calcium concentration in the cytosol of vascular cells is maintained between 0.1 and 0.4 µM by a very efficient homeostatic mechanism. This mechanism maintains a balance between the micromolar concentration in the cell and the millimolar concentration outside the cell, maintaining a steep concentration gradient across the membrane. This gradient is important for cell survival, and several systems are involved in its regulation (ATP-ases, translocases, channels and ion-carriers). The increase in calcium flux into the cell marks the first step of many toxic events.

Vitamin E has been shown to protect against calcium disequilibrium. Studies performed on isolated hepatocytes show that calcium-dependent toxicity is strongly related to the vitamin E content of the membranes [190]. Figure 5.2 illustrates a possible sequence of events leading from oxidative stress to cell lysis. Oxidative stress both increases membrane permeability and initiates peroxidation of the membrane. These effects may be related.With the depletion of glutathione and other antioxidants, the cell enters an irreversible phase that ends in cell lysis [148, 170].

The mechanism of cell lysis includes an interaction between a lipid peroxide (LOO˙) and vitamin E (TOC-H) with the formation of α-tocopherol free radical (TOC˙).

LOO˙ + TOC-H → LOOH + TOC˙
TOC˙ + GSH — enzyme → TOC-H + GS˙
2GS˙ → GSSG (oxidized glutathione) or
TOC˙ + vitamin C → TOC-H + vitamin C˙

Both regenerative pathways for vitamin E have been demonstrated experimentally. The drop in glutathione content of the cell is a common step in oxidation-induced lysis of the cell. The cellular content of glutathione is about 0.6 to 1.0 nmole/10^6 cells or 0.4 to 0.7 nmoles per mg protein. It is also estimated that one molecule of vitamin E protects about 500 molecules of membrane unsaturated fatty acids (in phospholipid). These estimates are supported by observations of the effect of treatment with anthracycline cytostatics (such as adriamycin) and the production of fatal cardiomyopathies [187, 206].

5.3 AMPLIFYING FACTORS

Free radicals are continuously produced in living organisms because of physiological activity: endogenous metabolism or detoxification of xenobiotics. Significant formation of reactive oxygen species may take place at the cellular level, or within tissues separately or simultaneously. The formation of free radicals in biological systems is self propagating and can undergo amplification.

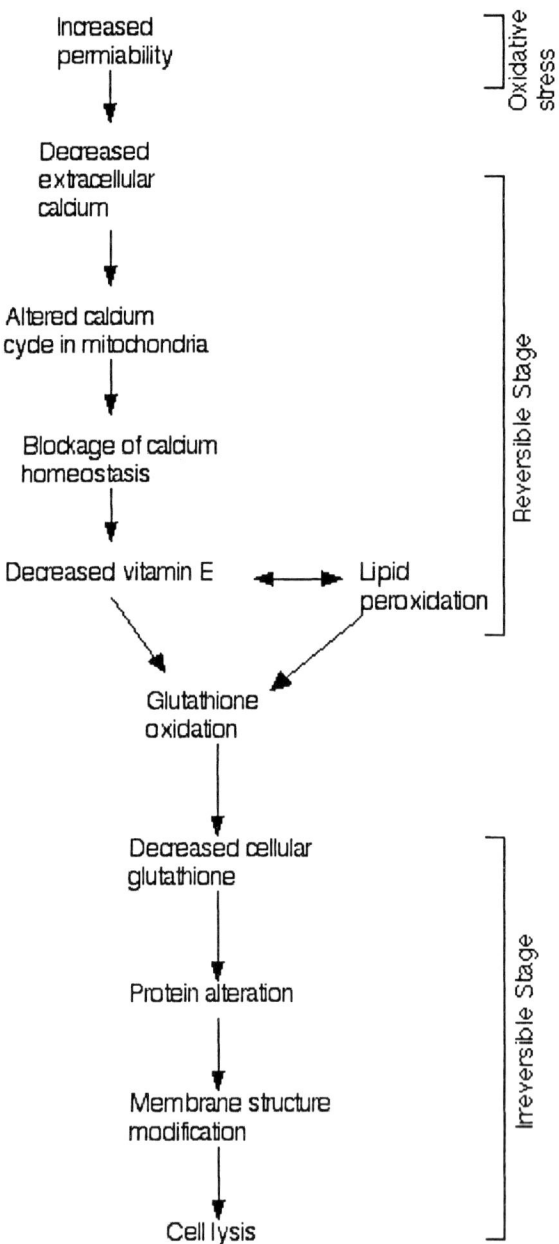

Figure 5.2. Perturbation of calcium homeostasis following oxidative stress.

During oxidative stress, local degenerative effects occur only if the antioxidant systems are overcome. This results in a primary lesion that may then spread (by propagation and amplification) to secondary sites and finally involve entire tissues (fig. 5.3). Once oxidative stress effects reach the tissue level, the effects are essentially irreversible and involve vascular modifications, ischemia, and cell death. This irreversible stage is not clear cut and depends on the organ and the body's ability to resist.

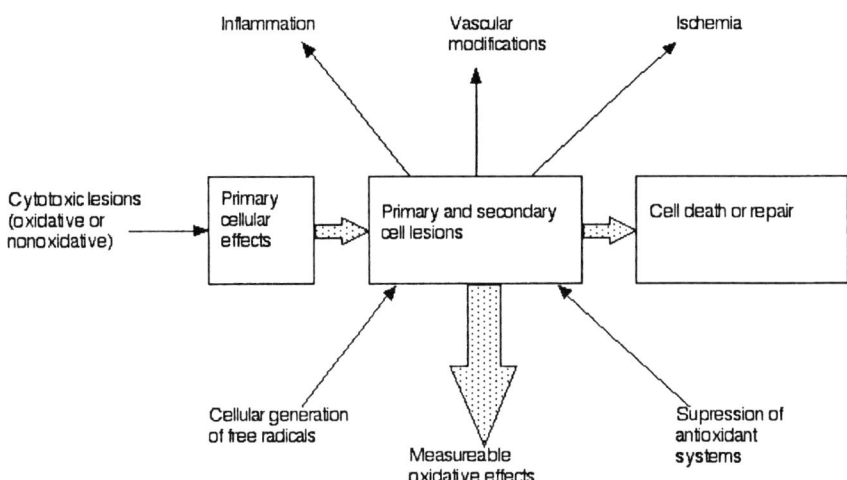

Figure 5.3. Scheme showing the differentiation of primary and secondary effects of lesions that appear following oxidative stress leading to the formation of reactive oxygen species.

Propagation of oxidative stress is a dynamic process that is complex and difficult to predict. It is heavily influenced by the balance between pro- and antioxidants in the system. Other mechanisms that aid in healing or recovery also come into play, which further complicates the prediction of outcomes. Phagocytic activity, inflammation, ischemia, altered membrane permeability, calcium decompartmentalization, protease activation, and other activities are involved as sources or consequences of oxidative stress. This complexity means the biochemical parameters that might reliably predict the outcome of oxidative stress on the organism are not understood. This is unfortunate as oxidative stress accompanies many diseases, such as chronic infections, and gastrointestinal, cardiovascular, and renal diseases. Some of these conditions will be considered in more detail in chapters 8 and 9.

5.3.1 Iron and Other Metallic Ions

The involvement of metallic ions (transition metals: Fe, Cu, Ni, Co, Cd) as prooxidants has been mentioned in earlier chapters. These ions accelerate the formation of the superoxide radical as well as the decomposition of lipid peroxides, which helps propagate the peroxidative chain reaction. These actions have been demonstrated *in vitro* using cell suspensions or tissue homogenates and take part in the Haber-Wiss reaction (see chapter 3.3). Halliwell and Gutteridge believe iron is the metal most commonly involved in these reactions *in vivo* [69, 72, 75]. However, while iron clearly participates in the Haber-Wiss reaction, not everyone who studies oxidative stress agrees it happens significantly *in vivo*. The involvement of iron in oxygen activation occurs in multiple phases of oxidative stress including nonenzymatic oxidation of glutathione and catecholamines, acceleration of peroxide decomposition, and promoting the formation of the very hazardous hydroxyl radical.

$$Fe^{2+} + H_2O_2 \rightarrow Fe^{3+} + OH^{\bullet} + OH^{-}$$

Irons ability to cause the decomposition of organic peroxides may place it in a central position in oxidative stress reactions in iron rich tissues, like muscle (fig. 5.4).

The controversy about the involvement of iron is related to the question of the extent of free ionic iron present in tissues. The organism is very effective in binding (chelating) free metal ions. This should not be surprising, given the toxicity of these free ions. Therefore, in plasma, almost all iron (about 100µM/dl) is bound to proteins like transferrin. When bound to transferrin, iron is not able to cause oxygen activation. The bacterialstatic capacity of plasma is based mainly on this molecule because iron is required for bacterial growth [103, 166].

Iron may be released from transferrin under special circumstances, particularly a decrease in pH. In spite of the homeostatic maintenance of a physiologic pH of 7.3, local decreases in pH may occur under inflammatory or hypoxic conditions [10, 189].

Another potential source of free iron is hemoglobin. In vitro studies have shown the small amounts of hemoglobin (10^{-7} to 10^{-5} M) acts as a prooxidant, favoring the peroxidation of polyunsaturated fatty acids. Therefore, open wounds in inflamed tissue or bleeding of the cornea offers favorable conditions for oxygen activation.

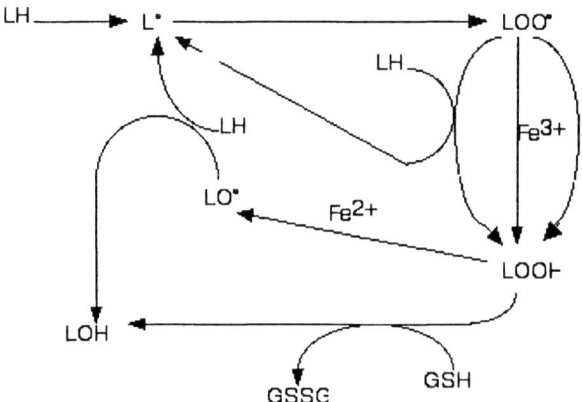

Figure 5.4. The amplifying role of iron in the peroxidation of lipids and the formation of hydroperoxides.

Similar situations may occur during oxidative stress in erythrocytes triggered by the presence of metallic ions such as lead. It should be remembered that, while free ionic iron is not present in plasma, it is present in cerebrospinal fluid (1.84±3 µM) and the concentration can increase to 2 to 3 µM in multiple sclerosis and epilepsy. In arthritis, the synovial fluid can contain more than 3 µM iron as well as lipid peroxides [72, 75].

A second controversy involves the prooxidative action of iron, which has a low reaction rate. Scientists introduce complexes of iron with oxygen, such as perferryl radical ($FeO_2^{\bullet}$) or the ferryl radical ($FeOO^{\bullet}$), to reactions to obtain a faster rate [69, 75]. Iron complexed with ADP, ATP, EDTA, or phosphate is also highly reactive. These complexes act especially well on lipid peroxides (LOOH) causing the formation of other radicals.

$$2\ LOOH \rightarrow LO^{\bullet} + LOO^{\bullet} + H_2O$$

The prooxidant efficiency of these iron complexes depends ont he iron/ligand ratio and the valence state of the iron. While ferritin is an iron protein that can bind and store up to 4,500 atoms of iron per molecule and has prooxidative activity, hemosiderin and lactoferrin are antioxidants.

There is, however, also evidence of prooxidant activity *in vivo*. Iron overloading takes place in hemochromatosis and transfusions, which can trigger several dysfunctions that can be corrected with chelating drugs such as desferrioxamine. Iron overload may produce dramatic effects in vitamin E deficiency or following intensive physical exercise [10, 55, 73]. The cytostatic drug, bleomycin, has a high affinity for binding iron. The resulting complex preferentially attacks DNA, producing cell death or genetic modifications [72, 75, 189].

In addition to iron, lead, cadmium, manganese, and cobalt increase lipid peroxides in blood plasma. These metal ions may be stored in brain and liver. They are normally isolated from the organism by being bound with albumin or metalothionein.

The role of iron as a promoter of peroxidation appears in some surprising situations. Vitamin C is a well known antioxidant. However, a system formed from ferric ions and vitamin C in low concentration is the best *in vitro* system for the production of the very dangerous hydroxyl radical. Does such a system function *in vivo*? The answer is that it may do so.

The enzyme heme oxygenase, mostly found in liver and spleen, catalyzes the opening of the hemoglobin ring, releasing free iron. The antimalarial drug, chloroquine, inhibits iron release from heme. Therefore, the induction of heme oxygenase may be appreciated as a cytoproductive response to destroy heme proteins (hemoglobin, myoglobin).

As mentioned before, a ferritin protein is capable of binding up to 4,500 atoms of iron. The protein itself, without iron, is called apoferritin. The intracellular generation of apoferritin constitutes a cytoprotective antioxidant system for endothelial cells. Indeed, it seems that all proliferating cells, including cancer cells, use it. In all forms of cancer, serum ferritin is elevated. This is particularly true during metastases. The excess ferritin is mostly generated in the tumor cells. Iron deficiency is known to slow the incidence of malignancy. Free iron promotes radical-induced DNA damage and mutations. Activating the iron within cancer cells to promote cell death is the mechanism of action of those cytostatic antibiotics (bleomycin) whose cytotoxicity depends on their formation of specific complexes with iron in tumor cells [136, 162].

5.3.2 Xanthine Oxidase

Xanthine oxidase has been called an enzyme that is full of mystery and surprises [78, 177]. Xanthine oxidase is a very efficient superoxide generating system. This system also produces chemiluminescence emission in the visible spectrum that is inhibited by catalase or ethanol, which inhibits the hydroxyl radical. The system also is able to promote the peroxidation of polyunsaturated fatty acids or lysosomes. During leukocyte activation, cytosolic xanthine oxidase activity is increased, producing significant amounts of superoxide [72, 123].

Whether in a cell free reaction or in a leukocyte suspension, the reaction between xanthine oxidase and its substrate, xanthine, produces large amounts of superoxide, with or without hydrogenperoxide. To what extent does this reaction occur *in vivo*? The single bit of evidence that it does occur is obtained from experimental ischemia. Allopurinol is a drug that inhibits xanthine oxidase, and is used in the treatment of gout because it prevents the formation of uric acid (the main product of the enzyme). Allopurinol also offers protection for animals with surgical ischemia. However, the xanthine oxidase activity found in blood is too low to account for the physiological effects it could theoretically cause in ischemia [66, 167].

The amplifying action of xanthine oxidase should occur because of tissue destruction and the release of purines (xanthine) from the nucleic acids. This type of tissue destruction occurs with some diseases involving inflammation or following treatment with cytostatics [74].

The mystery connected to xanthine oxidase comes first from its structure. This enzyme contains two irons, molybdenum, and FAD, and has a large molecular weight of 300,000. Therefore, it is possible that this large molecule has an internal redox cycle. In addition, it seems that the enzyme found in vivo is xanthine dehydrogenase, which oxidizes purines and pyrimidines but requires NAD. Xanthine dehydrogenase can easily be converted into xanthine oxidase, which performs the same reactions but with the formation of reactive oxygen species (superoxide and hydrogenperoxide). It is speculated that xanthine dehydrogenase first appeared in anaerobic bacteria and xanthine oxidase appeared later as a consequence of increasing oxygen levels [154].

Xanthine oxidase can oxidize at least 40 substances, including the purines xanthine and hypoxanthine, which are components of nucleic acids. Therefore, a certain relationship should exist between damage to nucleic acids and the activation of xanthine oxidase and the resulting uric acid production. Such a relationship seems to exist in cancer patients during treatment with cytostatics. Most of the released xanthine oxidase from plasma comes from destroyed leukocytes [142, 159].

The involvement of xanthine oxidase in aggravating oxidative stress seems to involve ischemia, when xanthine dehydrogenase is converted into xanthine oxidase, which then releases superoxide [159]. The conversion of xanthine dehydrogenase to xanthin oxidase is thought to be triggered by the accumulation of large amounts of xanthine that accumulate during ischemia because of the breakdown of ATP (fig. 5.5). This view is supported by experimental evidence. The use of allopurinol has produced only limited improvement and its use us continuing to be studied [66].

5.3.3 Bilirubin

Bilirubin is the final product of hemoglobin degradation (Hb = hemoglobin, Heme-OH = hemehydroxylate).

$Hb\text{-}Fe^{2+}$ — cytochrome P_{450} → $Heme\text{-}OH\text{-}Fe^{3+} + O_2^{\bullet}$
$Heme\text{-}OH\text{-}Fe^{3+}$ — heme oxygenase → biliverdin + Fe^{3+}
biliverdin + NADPH — cytochrome C reductase → bilirubin

These enzymatic reactions take place in the liver. Bilirubin is toxic and therefore circulates in the plasma mostly bound to albumin. Bilirubin is lipophilic and destroys cellular membranes by reacting with phospholipids (phosphatidylcholine). It crosses the blood-brain barrier, inhibits oxidative phosphorylation, and decreases cAMP concentration in tissues.

However, experimental evidence from cell free systems shows that bilirubin is a strong antioxidant through the scavenging of singlet oxygen. Evidence of a beneficial in vivo antioxidant role for bilirubin is yet to be established.

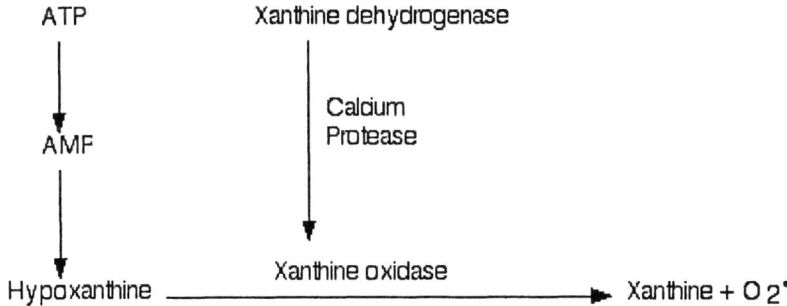

Figure 5.5. Proposed mechanism for the involvement of xanthine oxidase in the formation of reactive oxygen species. Conversion of xanthine dehydrogenase to xanthine oxidase is facilitated by ischemia.

During chemical damage to the liver (table 5.4) an increase in heme oxygenase and lipid peroxidation is proportionally related to the drop in glutathione [6, 20, 27, 126, 128]. Modifications to liver function seem to have a threshold that is mostly related to the glutathione concentration. Only after crossing this threshold of glutathione concentration does lipid peroxidation and bilirubin production increase.

Table 5.4 Biochemical modifications in the liver of rats after chemical exposure [126].

Condition	Peroxides (nmol MDA/mg prot.)	Hemoxygenase (nmol bilirubin/mg prot.)	Glutathione (µmol SH/mg prot.)
Control	3.85±0.12	1.68±0.23	6.16±0.22
$CdCl_2$ (2.5 mg/kg)	6.35±0.15†	5.87±0.13†	3.46±0.18‡
APH* (50 mg/kg)	6.39±0.19†	3.24±0.17‡	4.32±0.42‡
APH* (100 mg/kg)	9.85±0.24†	6.58±0.34†	3.12±0.25
Ethanol (80 mg/kg for 2 months)	4.28±0.25	2.47±0.25	3.82±0.55
Ethanol (80 mg/kg for 4 months)	10.17±0.34†	5.68±0.18†	0.95±0.43†

*APH = Acetylphenyl hydrazine
†$p<0.01$
‡$p<0.05$

A similar threshold effect has been shown in humans (table 5.5). Again, there are significant increases in peroxides and bilirubin when glutathione decreases. It seems that in the liver, the drop in glutathione allows the increase in lipid peroxides and bilirubin. Therefore, the increased formation of bilirubin matches the extent of liver failure and the increased formation of lipid peroxides. In newborns with jaundice, the increased level of bilirubin is related to an increased level of lipid peroxides. Newborns with hyperbilirubinemia also have an increased incidence of encephalitis. The linear correlation between lipid peroxides and bilirubinemia is 0.76 and 0.97 for hepatic coma [128].

In spite of these clinical studies, *in vitro* experiments using liposomes or cell free systems have suggested that bilirubin is an antioxidant. The *in vivo* formation of bilirubin free radicals has been demonstrated during phototherapy studies as newborns with jaundice are treated with exposure to the sun or an artificial UV source. It seems that both situations may be true. Bilirubin, like ascorbate, is able to function as a pro- or antioxidant. As Ames noticed, the antioxidant capacity of bilirubin is higher when bound to albumin [5]. In liver failure, as albumin concentration is decreased, more bilirubin is found in the free, toxic, state.

Table 5.5 Biochemical modifications in blood of patients with liver deficiencies [126].

Condition	n	Peroxides (µmol/l)	Bilirubin (mg/dl)	Glutathione (µmol SH/ ml)
Control	30	2.5±2.64	1.02±0.68	43.52±5.83
Chronic hepatitis	40	9.86±2.52*	4.36±0.12†	45.46±6.21
Viral hepatitis	25	23.41±6.56†	10.74±1.24†	33.12±4.34*
Ethylic hepatitis	15	21.84±7.31†	7.43±0.75†	28.67±5.62†
Cirrhosis	10	18.53±5.64†	14.62±5.24†	24.62±4.72†
Hepatic coma	5	68.51±8.33†	22.74±8.63†	23.43±4.97†

*$p<0.05$
†$p<0.01$

Chapter 6

RADIATION AND FREE RADICALS

6.1 UV RADIATION

UV radiation causes damage by the process of phototoxicity. These effects can be harnessed as phototherapy for the treatment of psoriasis and other conditions.

6.1.1 Photochemical and Photosensitivity Reactions

Photochemistry includes a large group of reactions triggered by the exposure to sunlight, or more precisely to the UV component of sunlight. These reactions are widespread, occurring in plants (photosynthesis), industrial situations (photography, bleaching), and directly on people (tanning, sunburn). UV action occurs first by transferring photons of energy to molecules with an appropriate structure and electronic configuration. According to Einstein's second law of photochemistry, the absorption of a photon triggers an activation of the molecule that reacts by changing its structure or by reacting with other molecules, mainly oxygen.

Photosensitive reactions are more complex. These reactions imply the existence of special electronic configurations called triplets or singlets that last 10^{-3} to 10^{-6} seconds. Such reactions involve structural changes and are illustrated in figure 6.1A.

Excited electronic states are unstable and the molecule returns quickly to its fundamental electronic state by releasing energy as heat or light (fluorescence). In suitable molecules, such reactions produce stable modifications (isomers, ionizations, or free radicals). Some molecules that exist in the excited state will react with other molecules in biological systems resulting in modifications and lesions (fig. 6.1B). Once free radicals are formed, their reaction with oxygen is unavoidable (see chapters 1.3 and 3.2). The resulting toxic effects depend on the nature of the biological substance (proteins, nucleic acids) and on the light intensity and wavelength.

Photosensitive substances, following their exposure to UV light are activated to the excited state with the formation of a triplet. The triplet state lasts long enough for those molecules to donate their energy to other molecules. Photosensitive substances include dyes, such as methylene blue, eosin, acridine orange; drugs, such as chlorpromazine; and natural biological compounds, such as protoporphyrin. The mechanism of photosensitive reactions depends on the molecule's structure and that of any substrates and the oxygen concentration.

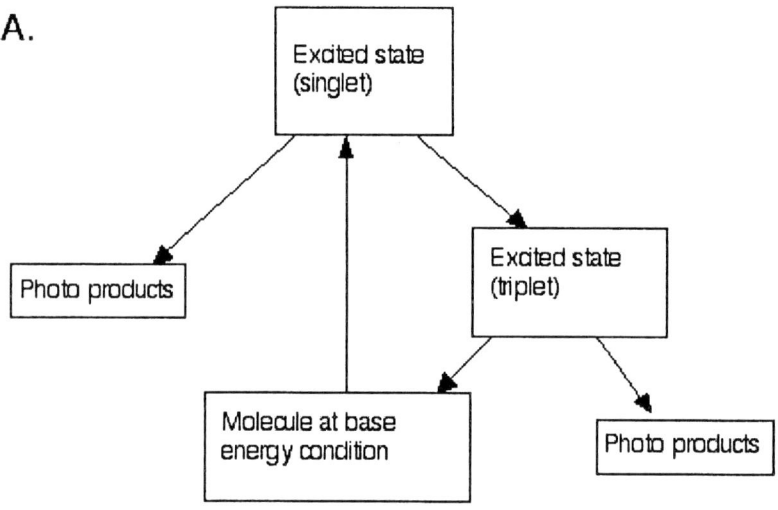

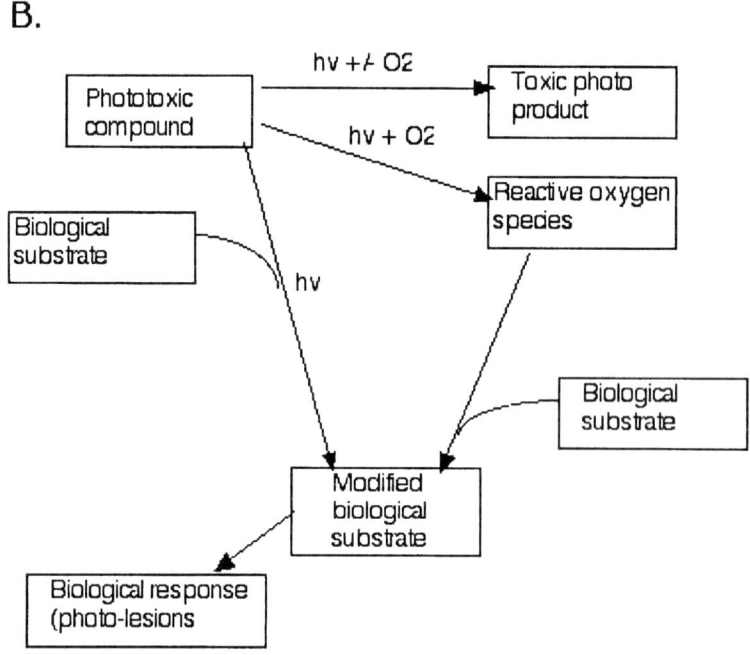

Figure 6.1. Photochemical reactions. A, Movement of excited electrons by photo energy; B, The dynamic involving a phototoxic compound and the production of biological effects.

Photosensitive reactions involve the transfer of electrons or hydrogen atoms (^3S = triplet state substance, RH = reacting substance, $^1O_2$ = singlet oxygen).
^3S + RH — electron transfer   S + R (Type I reaction)

$^3S + O_2$ — electron transfer $S + {}^1O_2$ (Type II reaction)

A photodynamic effect results following the action of a photosensitive compound. This effect is seen upon exposure to UV in any organism, from bacteria to human patients. These effects are the result of the effect of UV radiation on amino acids, proteins, or nucleic acids. Therefore, phototoxic effects are initiated at the molecular level and then extend to the entire organism. These final effects include irritation, itching, skin lesions, and necrosis. A phototoxic effect quickly runs its course, lasting about 24 hours. Photoallergic effects, on the other hand, are long lasting. These reactions produce photoallergens that act as haptenes as a consequence of protein structural modifications.

Like all free radical induced reactions, photosensitive reactions begin at the molecular level. The main biological substrate is formed by the aromatic amino acids, tryptophan, tyrosine, and phenylalanine, and sometimes methionine or cysteine. The principle characteristic of these amino acids is their absorption of light at wavelengths between 2,500 and 2,900 Å because of the aromatic ring. Tryptophan represents a special case. This amino acid is easily oxidized, producing singlet oxygen and formyl-kynurenine. Most photochemical studies have shown that tryptophan is the principle target of photo action on proteins and hydrogenperoxide is a by-product [123, 198]. Amino acids in a photosensitive reaction may be coupled with photosensitive dyes or porphyrins. Polyunsaturated fatty acids might also be coupled with these photosensitive reactions, resulting in peroxidation. The combination formyl kynurenine and hydrogenperoxide seems to be highly toxic. It damages DNA, producing scissions of the double helix, and destroys phages and microbes [51, 198].

A large number of animal and plant cells, both procaryotic and eucaryotic, may have respiration or growth inhibited following exposure to sunlight or UV with a wavelength longer than 300 nm. Cell death is a consequence of photoproducts and lipid peroxidation in cell membranes. Therefore, photo action on organisms is a double edged sword. Light is necessary to support life, but photo induced reactive oxygen species are also produced that cause damage.

6.1.2 Photochemicals and Ozone

Thanks to space science's interest in the atmosphere, the existence of unpredictable photochemical reactions in the atmosphere and the involvement of free radicals is now well established. In fact, most atmospheric pollutants favor the production of free radicals with photochemical or biological consequences.

In order to understand the possibility of the presence of free radicals in the atmosphere, we should review some basic knowledge. The sun emits approximately 4 x 10^{23} KW/s. Sunlight on earth amounts to about 2 calories per minute per square centimeter (1,400 V/m²). The atmosphere's nominal composition (78.08% nitrogen, 20.95% oxygen) varies with geographic area, season, and pollution. A vast array of reactions takes place between the normal components of the atmosphere. Chemical pollutants, when added to this mix, result in unpredictable oxidative photochemical reactions.

In spite of the complexity of these reactions, some general information is known (Table 6.1). Several facts appear from the reactions described in table 6.1. The first five reactions are pure photochemical reactions that proceed at a high rate. The remainder have a slower reaction rate and are catalyzed by UV radiation. Secondly, these reactions show that free radicals can exist under the conditions present in the atmosphere. The reactions in the table also point out that ozone is a major participant in these reactions.

Table 6.1 The main photochemical reactions that occur in the atmosphere [99].

Reaction	Rate constant $(M^{-1}s^{-1})$
$O_3 \xrightarrow{h\nu} O^{\bullet} + O_2$	1.6×10^{-5}
$O_3 \xrightarrow{h\nu} O + O_2$	3.6×10^{-4}
$NO_2 \xrightarrow{h\nu} NO + O$	5.6×10^{-3}
$H_2O_2 \xrightarrow{h\nu} 2\, OH^{\bullet}$	4.6×10^{-6}
$CH_2O + 2\, O_2 \xrightarrow{h\nu} 2\, HO_2^{\bullet} + CO$	1.7×10^{-5}
$O_3 + NO \xrightarrow{h\nu} NO_2 + O_2$	2.0×10^{-12}
$O_3 + HO_2^{\bullet} \xrightarrow{h\nu} OH^{\bullet} + 2\, O_2$	1.6×10^{-12}
$HO_2^{\bullet} + NO \xrightarrow{h\nu} NO_2 + OH^{\bullet}$	3.7×10^{-12}

The ozone layer is found between 25 and 30 Km of altitude. This sometimes decreases to 12 Km above the earth's magnetic poles. It is well accepted that life could not have appeared without the protective effects of the ozone layer, which absorbs large amounts of UV radiation, especially the most harmful wavelengths of 2,600 to 3,800 Å. The 40% decrease in the ozone layer above Antarctica (the ozone hole) seems to be caused by the release and accumulation of chlorflurocarbons once widely used in refrigeration and the cosmetics industry. A large number of papers have predicted an increase in skin cancer as a result of higher UV exposure. The direct effect of UV radiation (especially UV A) in the etiology of skin cancer is well documented both experimentally and epidemiologically [127, 198]. The potential effect of variation in atmospheric ozone concentration on skin cancer is not entirely clear. Ozone does seem to have a natural variation that is related to the sun spot cycle. There is also some evidence that the frequency of some diseases, including the mortality of heart disease patients is influenced by these changes [127, 150].

Ozone, however, can have other impacts on human health. Ozone reacts readily with hydrocarbons under laboratory conditions. Normally the reaction between ozone and hydrocarbons is slow (K = 10^3 to 10^5 $M^{-1}s^{-1}$), but in the atmosphere, by coupling with photochemical reactions, the process is accelerated and involves the formation of free radical and peroxides. The result is photochemical smog that is often found over large population areas like Los Angles or Tokyo.

$NO_2 - \lambda < 430$ nm $NO + {}^1O_2$
hydrocarbons + ${}^1O_2$ organic peroxides

For example: CH_3-CH-CH_3 (propane) + O_3 CH_3-CH(OOH)-CHO (peroxide)
ozonides $2 CH_2O + HCOOH$

The result is the production of formaldehyde, peroxide, and formic acid, all irritating compounds. Traces of metals, such as lead, aluminum, or copper in the atmosphere accelerates these reactions.

The formation of photochemical smog is a major factor in the etiology of several respiratory diseases and eye irritation. In these effects, the photochemical reactions involved are due to the participation of chemical pollutants that result from the burning of fossil fuels (coal, petroleum) and to special meteorologic conditions.

As seen in table 6.2, the formation of smog is variable, and depends on local conditions. The recorded ranges of composition are (in parts per billion) CO_2, 200 to 2,000; nitrogen oxides, 5 to 25; hydrocarbons, 20 to 50; ozone, 2 to 20; aldehydes, 5 to 25; peroxyl nitrates, 1 to 4 [99, 127]. Peroxyl nitrates (PAN), having the general structure R-CO-$OONO_2$, are stable in the atmosphere and are very irritating to the eyes. PAN results from the following photochemical reactions occurring in the atmosphere.

$CH_2O - h\nu$ $HCO^{\bullet} - O_2$ $HCOOO - NO$ $HCOONO_2$ (PAN)

Peroxiproprionyl nitrate (CH_3-CH_2-CO-$OONO_2$) is a PAN that is lethal for rats at a concentration of 108 ppm and also damages plants by destroying glutathione and SH-containing enzymes, such a glucose-6-phosphate dehydrogenase. PAN also oxidizes indoleacetic acid, NAHPH, and olefins.

PAN + olefins   epoxides + $CH_3COO^{\bullet}$ + NO_2

The reactions summarized here are actually only a small portion of the many, complex photochemical reactions that take place in the atmosphere. The knowledge of the reaction mechanisms may be useful to reduce noxious emissions leading to chemical pollution. It should be mentioned that due to the great affinity of UV radiation and ozone for polyunsaturated fatty acids, plants and insects undergoing long term exposure have a reduced fatty acid content. But aquatic organisms, which are less exposed to atmospheric chemical pollutants, are less affected [41, 94, 127].

Table 6.2 A comparison of the characteristics of photochemical smog in London and Los Angles.

Characteristic	London	Los Angles
Peak period	Morning	Noon
Optimal temperature	-3 to 60 C	23 to 340 C
Humidity	High with fog	Low with clear skys
Thermal condition	Inversion	Overheating
Chemical condition	Reduction	Oxidation
Principle effect	Respiratory	Ocular
Main component	Sulphur oxides	Hydrocarbons and nitrogen oxides

6.1.3 Photodynamic and Phototoxic Effects

The effects of damaging wavelengths of UV are summarized in table 6.3. UV-B, which comprise about 0.2% of all UV radiation that reaches the earth, is the most dangerous. In order to produce harmful effects, UV radiation must penetrate into the dermis (stratum corneum). There it can act on phospholipids, proteins, and nucleic acids. These phototoxic effects are characterized by tardive erythema (sunburn), followed by hyperpigmentation (tanning), and desquamation (peeling). To differentiate normal tanning from phototoxic effects, the toxic effects are characterized histologically by denatured epidermal cells and necrosis. This leads to the early aging of skin (actinic elastosis) seen in people constantly exposed to the sun (farmers, sailors, etc.).

Table 6.3 Direct effects of UV radiation classes.

UV-A (280 - 315 nm)	UV-B (315 - 400 nm)	UV-C (280 nm)
• Tanning • Erythemia	• DNA damage • Inhibition of DNA biosynthesis • Synthesis of vitamin D • Late effect erythemia • Skin cancer	• Late effect erythemia • Structural alterations of nucleic acids • Photosensitivity • Skin cancer

In spite of their frequency, phototoxic effects on the whole are not as dangerous as photoallergic effects. These effects are more complex, consisting of a wide range of responses of the body, including the longer incubation, humoral type of immunity. The most common and persistent morphologic changes are hypertrophy of melanosomes, which are dermal cells that are rich in melanin pigment and favor the formation of dangerous melanoma.

The photoallergic effect is produced by photosensitivity reactions that occur at lower UV energies than phototoxic effects. The most common clinical signs consist of urticaria, morphological modifications of papular and eczematous cells.

Porphyria is a typical photosensitive reaction of the organism triggered by protoporphyrin. This condition may be congenital or acquired (such as by lead poisoning). During acute attacks, the patient with erythtopoetic porphyria develops edema and erythema following exposure to sunlight. Their blood contains increased lipid peroxides due to increased erythrocyte sensitivity to oxidation. Lipid extracts from erythrocyte membranes of these patients are able to lyse the red blood cells of healthy people.

Newborn jaundice results in increased amounts of lipid peroxide in the blood [123, 203]. Bilirubin accumulates due to a temporary deficiency in glucuronyl transferase, the enzyme that catalyses the conversion of liposoluble bilirubin into a water soluble glucuronylated derivative. The production of lipid peroxides in the blood is due to the insufficient development of enzymatic antioxidant systems. Newborn jaundice is commonly treated by exposure to the sun or another source of UV light. Bilirubin is destroyed by the singlet oxygen produced and the end products are harmless.

Photo-induced cancer has been extensively studied, and free radicals have been shown to be involved [112, 137, 198]. In the etiology of skin cancer, the direct action of UV (table 6.3) is involved as well as the photosensitizing action of some chemical pollutants (polyaromatic hydrocarbons and tyrosine) on nucleic acids, producing structural modifications. Other, indirect, mechanisms include N-formyl kynurenine production from tryptophan photolysis, which is converted enzymatically to carcinogenic compounds (figure 6.2), and cholesterol epoxide that results from cholesterol hydroperoxide.

There is a well established relationship between UV exposure, skin cancer incidence, and the antioxidant content of the skin. As in chemically-induced tumors, an incubation period occurs. During this phase, low level biochemical modifications may take place leading to tumor production, or homeostatic mechanisms may return the system to normal. According to Miller's hypothesis, most carcinogenic compounds are produced as a consequence of metabolic activation [12]. For most skin cancers, the etiology may involve photo conversion of procarcinogenic compounds into active carcinogens. These active carcinogens may be a variety of compounds, including peroxide derivatives, resulting from UV action that then act on DNA, producing structural modifications (such as thymine dimerization). The protective role of antioxidant supplementation against skin cancer (either chemically- or UV-induced) has been demonstrated [4, 59, 71].

Melanoma is, perhaps, the most aggressive form of cancer known, and , as is shown by its name, is an anarchic development of melanogenesis. Melanogenesis is a physiological process that produces the melanin pigments that are necessary for the color of hair, the skin, the iris, and substantia nigra of the brain. The main role of melanin seems to be as a photoprotector. Melanin's structure is not clear because it is an insoluble polymer that contains quinone and quinone-related derivatives. Melanin pigments also contain stable free radicals. Melanin pigments may be classified according to their color and source as:

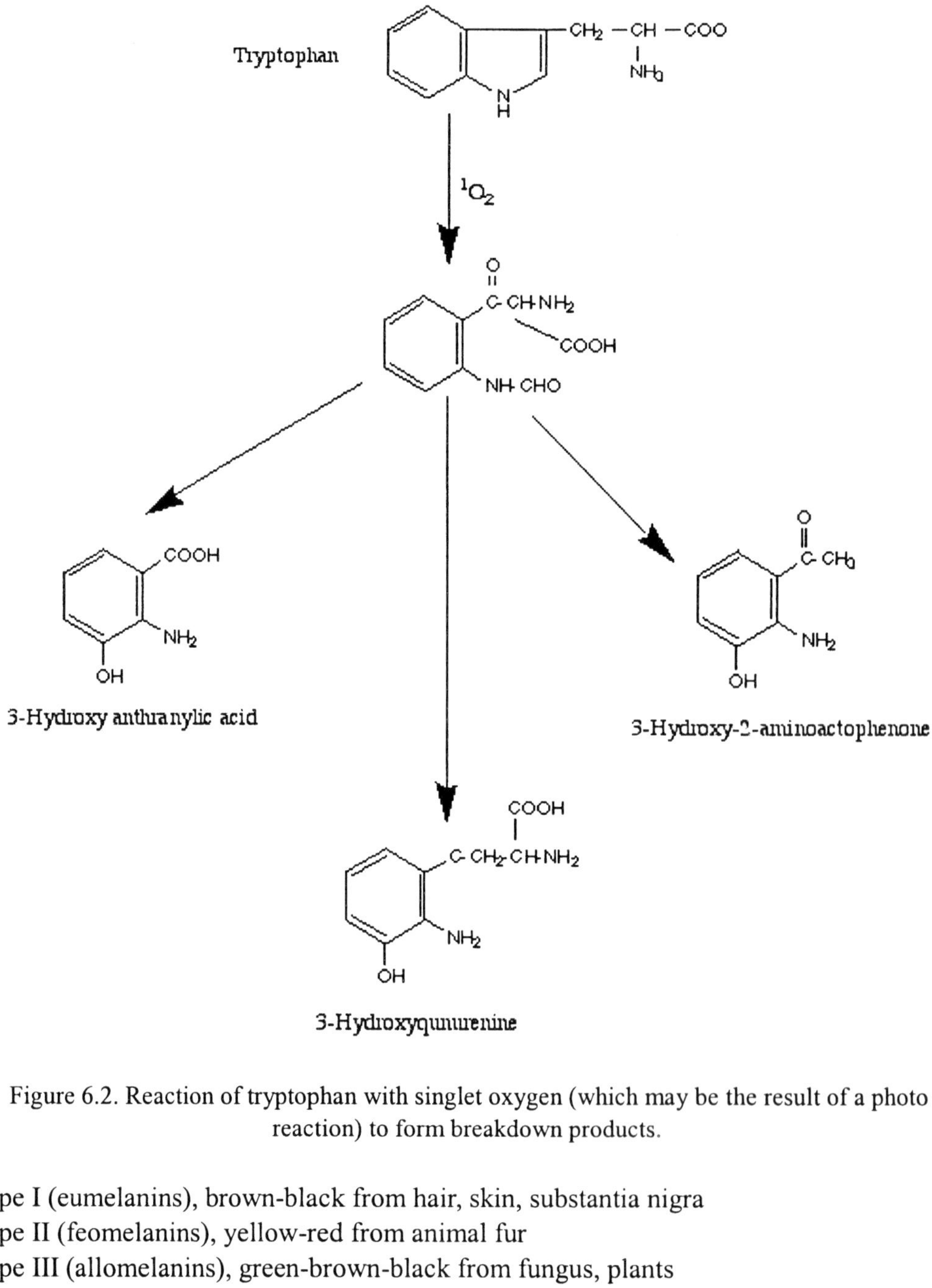

Figure 6.2. Reaction of tryptophan with singlet oxygen (which may be the result of a photo reaction) to form breakdown products.

Type I (eumelanins), brown-black from hair, skin, substantia nigra
Type II (feomelanins), yellow-red from animal fur
Type III (allomelanins), green-brown-black from fungus, plants

Melanin pigments are produced by an oxidative chain of reactions that are enzymatically mediated. Therefore, type I melanins arise from the amino acid tyrosine or from dihydroxyphenylamine (DOPA). Type II arise from DOPA and cysteine. Type II arise from catechols and various aromatic reactions. These precursors undergo successive oxidations and polymerizations, in the process of which free radicals are produced. These free radicals may remain in a free state or a quinones and semiquinones. The free radical concentration varies between 10^{-17} to 10^{-19} spins/g dry weight or about one free radical per monomer having a molecular weight of 150 [72, 114, 127]. Melanin pigments readily absorb water, ions, and small proteins (20% to 50% of their weight). Following exposure to light, melanins undergo photolysis with the formation of superoxide and hydrogenperoxide as well as singlet oxygen. When all the components are coupled, a very complex structure is formed. Thus, melanins are peculiar compounds. The photoprotective and antioxidant properties of melanins are actually a double edged sword, having both beneficial and dangerous effects. The production of free radicals in these dangerous effects is a major factor.

Phototherapy is a potentially beneficial action of UV radiation. It has been mentioned in the treatment of newborn jaundice. The skin lesions produced by Herpes simplex virus are also treated by photosensitive reactions using red neutral dye. Thirty years ago, a phototherapy for skin cancer was attempted using dyes such as hematoporphyrin or acridine orange.

In the USA and Western Europe, limited success has been had using coal tar and UV-A (2,900 to 3,200 Å) to treat psoriasis [113, 140]. These phototheraputic treatments were criticized as possibly leading to skin photosensitization or the amplification of skin cancers due to photo effects. Drugs based on psoralens (furocoumarins) are now commonly used to treat psoriasis. Clearly, considering the effect of UV radiation on nucleic acids or phospholipids, phototherapy has some risk. However, restriction of the UV exposure to the A band minimizes the risk.

6.1.4 Photosensitizing Drugs

The enormous increase if drug treatment has allowed the discovery of adverse side effects. Among these side effects, some photoallergies have appeared when some drugs are use for long periods of time (table 6.4).

The principle reaction of these drugs consists of their (or a metabolite's) interaction or combination with proteins, resulting in an antibody (covalent adduct) that produced clinical consequences by acting on activated lymphocytes.

Drug + protein — $h\nu$ antibody   erythema, edema, etc.

The respective antibodies result from drug metabolites or their free radical intermediates. Under the action of UV radiation, differentiation may occur.

Type I: drug + UV-A   antibody + oxidized protein
Type II: drug + endogenous photosensitization   oxidized protein

Oxidized proteins may affect the immune system by triggering cutaneous eruptions. Type I reactions can occur with 3,3,4,5-tetrachlorosalicyl-anilid, while type II reactions can occur with quinine. Quinine, following extended use, is metabolized to a by-product that will react with tryptophan, producing singlet oxygen and the endogenous photosensitivity derivative, N-formyl kynurenine.

These side reactions have been demonstrated for chlorpromazine and psoralens, whose metabolites react with DNA (pyrimidine bases) resulting in covalent adducts. As was later found, these drugs also affect membrane phospholipids, but without forming peroxides. Rather, only singlet oxygen is produced.

Table 6.4 Drugs known to produce photoallegic reactions.

- Psoralens (furocoumarins)
- Phenothiazines (chlorpromazine)
- Quinine
- Protriptilyn
- Tertacyclin antibiotics
- Sulfonamide
- Nalidixic acid
- Halogenated salicylanilides

6.2 IONIZING RADIATION

6.2.1 Effects of Radiation at the Cellular Level

Because of the military implications, many studies of the effect of ionizing radiation on man were carried out in the 1950s and 1960s. It was during this period most of our knowledge of the effect of ionizing radiation on organisms was learned. Radiobiology was the first biological field in which free radicals were found in living organisms. This was caused by exposure to ionizing radiation, and radiation sickness was the first disease in which free radicals were clearly demonstrated to be involved. Studies during this time also showed that the involvement of free radicals begins at the molecular level and then progresses to the cellular level.

The environment in which living organisms reside has abundant sources of ionizing and other forms of radiation. Humans are exposed every hour to about 10^5 neutrons, 10^5 cosmic rays, 3×10^4 atoms of Rb, Bi, Po, and Pb from the air, 15×10^6 atoms of K^{40}, 7×10^3 atoms of uranium from food, and 2×10^8 photons [197]. Over evolutionary time, organisms adapted to this environment by including protective and repair systems in their biological make-up. It was possible for these systems to evolve because the exposure to these radiations and their effect is low, and is made lower because most do not penetrate into the organisms tissues. The mean yearly dose from natural radioactivity is approximately 1,870 µSv [197, 198]. Ionizing radiation that penetrates the organism's body produces biological consequences as a function of their energy. Thus, a particle that penetrates the body could produce the excitation of an atom (requiring 4 eV) or its ionization (requiring 33 eV). If free radicals are to be produced, higher energies are required. Therefore, alpha particles and neutrons are the most dangerous as they produce a large number of ionizations during their passage.

Ionizing radiation produce damage directly by the interaction of particles on cells and tissues or indirectly through the activation of oxygen. For gamma radiation, both effects occur in equal amounts. For neutrons the direct effect is predominant.

The breaking of the water molecule by ionizing radiation (radiolysis) produces many free radicals and ions (H_2, H_2O_2, e^-aq, $H^{\bullet}$, $HO_2^{\bullet}$, $OH^{\bullet}$, H_3O), most of which are reactive oxygen species. The amount of free radicals (reactive oxygen species) produced depends on the type and energy of the ionizing radiation and the type of organs involved. Thus, irradiation with alpha particles produce high linear energy transfer values because the free radicals formed are capable of propagating and producing tissue damage.

As described in greater detail in chapter 2, when energy is absorbed by the organism, the activation of oxygen occurs, producing reactive oxygen species and their secondary products.

$$H_2O_2 + e^-aq \quad OH^{\bullet} + OH^-$$
$$H_2O_2 + H^{\bullet} \quad OH^{\bullet} + H_2O$$

Fifty years after the discovery of water radiolysis, the hydroxyl radical ($OH^{\bullet}$) is now recognized as the most damaging product in the indirect action of ionizing radiation. This radical is capable of destroying any organic molecule and preferentially attacks cell membranes and the sulfhydryl groups of proteins and nucleic acids.

While damaged proteins can be replaced by the body with minimal or no consequences, damage to DNA may result in mutations and cancer. Ionizing radiation damages DNA in different ways resulting in modification to purine and pyrimidine bases, peroxidized ribose, new product formation (such as 5-hydroxyuracil, 8-hydrodeoxyguanasine, thyminglycol). The quantitation of such by-products in a person's urine is a very effective marker radiation sickness as has been well demonstrated after accidental high-dose nuclear exposure [5, 197]. However, the same products can be detected (at much lower concentrations) after exposure to natural radiation, following oxidative stress, or as a consequence of aging.

6.2.2 Radiation Sickness

Radiation sickness involves clinical and biochemical modifications that appear immediately or shortly after exposure to ionizing radiation. When the exposure is low or there is a low dose rate, the consequences at the molecular level are minor because the damage is quickly repaired (see figure 6.3). At high doses or a high dose rate, the biological consequences are severe and in extreme cases can result in death in a few days. At high doses, the organism's protective and repair systems are overwhelmed and the ionization-induced modifications that began at the molecular level extend to the cellular level. Damage progresses to produce tissue damage and then the entire body exhibits clinical evidence of the injury. The free radicals produced upon exposure to the ionizing radiation are responsible for most of these progressive effects.

The DNA may be considered the target of ionizing radiation because damaged proteins have a greater possibility of being replaced before permanent damage occurs. If the damage to DNA is extensive enough, however, repair cannot occur before serious adverse events beginning to unfold. These events generally involve fatal cancers. This can also occur with long-term (up to 10 years or more) exposure to ionizing radiation. Therefore, it is necessary to differentiate acute radiation sickness from chronic radiation exposure that does not lead to disease.

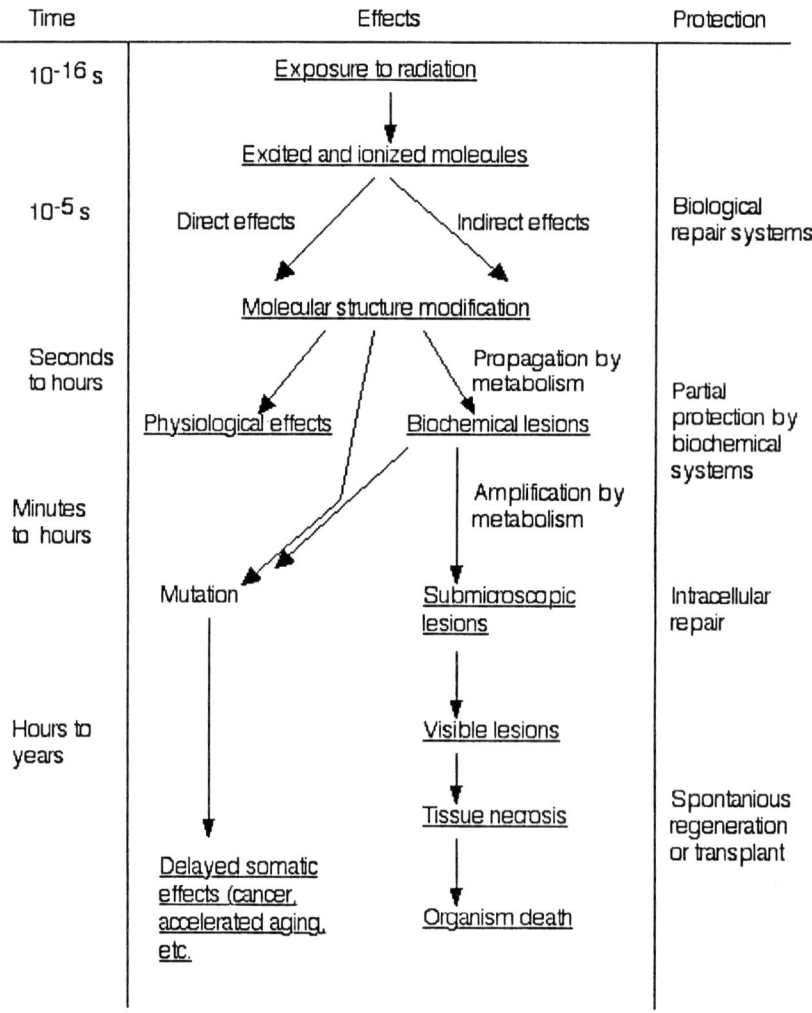

Figure 6.3. Principle biological modifications produced by ionizing radiation, the times involved, and possible protective activity.

As seen in table 6.5, the clinical symptoms of radiation sickness depend on the dose of radiation received. The dose also affects the biological consequence (or syndrome). Different tissues differ in their sensitivity to ionizing radiation. Lymphocytes and bone marrow stem cells are the most radiosensitive cells in the organism [198]. This becomes apparent in table 6.5 as the dose differences clearly affect different parts of the body. Eighty years ago, Bergonie and Tribondeau explained that cellular radiosensitivity is directly related to the tissues rate of cell division (growth) and inversely related to the degree of functional and morphological differentiation of the cells. Cells are sensitive to radiation during the G_1 (presynthetic) period of mitosis.

Table 6.5 Clinical signs of acute radiation sickness.

Dose (Gy)	Frequency (%)	Effects	Clinical signs	Critical period	Mechanism	Mortality (%)
0 to 1	10	Reversible	Lymphopenia	---	---	0
1 to 2	30	Reversible	---	3 hours	---	0
2 to 5	50 to 90	Hematologic	Leukopenia Hemorrhage Infections	2 to 6 weeks	Cellular disturbances, necrosis	0 to 90
5 to 15	100	Intestinal	Diarrhea Fever Electrolite imbalance	3 to 14 days	Shock	90 to 100
16 to 50	100	Cerebral	Diarrhea Convulsions Coma	1 to 48 hours	Cerebral edema	100

6.2.3 Oxygen Effects and Therapeutic Irradiation

As far back as 60 years ago it was observed that the biological effects of ionizing radiation are diminished in hypoxic conditions. Later, experimental evidence confirmed this observation and showed that this decrease in tissue damage occurs only if the partial pressure of oxygen at the moment of exposure is low. This oxygen effect is only observed with X-rays and gamma rays. Neutrons are not influence by the oxygen partial pressure [197].

The oxygen effect on the action of ionizing radiation was not understood until the mechanism of oxygen activation was discovered. Oxygen activation and the formation of reactive oxygen species fully explains the oxygen effect of ionizing radiation. Irradiation of organisms under high oxygen partial pressures enhances the effect of the radiation as more reactive oxygen species are produced. The oxygen effect was also demonstrated in cell-free media through the irradiation of aerated aqueous solutions. A direct, proportional relationship was found between the radiation dose and the concentration of hydrogenperoxide formed by water radiolysis.

The extent of exposure of DNA to the effects of ionizing radiation may be illustrated by the following calculations. Given that cells have a diameter of 100 to 300 Å, it follows that 10^9 cells occupy a volume of 10 microns square with weight of 1 gram. If one roentgen (0.01 Gy) produces 2×10^{12} pairs of ions per gram tissue, than 1 Gy will produce 10^6 pairs of ions per cell. Since the DNA of the cell occupies about 10% of the cell's volume, 10^5 pairs of ions will reach the DNA [197].

The use of ionizing radiation to treat cancer is based on the fact that cancerous cells undergo an increased rate of mitosis and, therefore, should be more sensitive to radiation. This leads to the paradox that radiation (in high doses) causes cancer, but, when given in moderate doses to a localized area, it may be capable of limiting tumor growth. The success of the use of ionizing radiation in the treatment of cancer depends on the dose used. High doses have been used to destroy as much tumor tissue as possible. However, this presents a double edged sword as radiation damage may be initiated in surrounding tissues during the treatment. Side effects, such as radiodermatitis, decreased immunologic function, and alopecia, are also common. In addition, solid tumors are hypoxic compared to surrounding tissues, which means a greater radiation dose is needed to have an effect. To alleviate the problem of hypoxia, radiosensitizers such as nitroimidazol, metronidasol (Flagyl), or misonidasole have been used, but with limited success. The use of radiosensitizers is also limited by the side effects and by the potential lack of affinity of the drug for the tumor.

Following the therapeutic use of ionizing radiation when using a radiosensitizer, anionic free radicals are produced.

$$RNO_2 + e^- \quad RNO_2^{\cdot} \text{ (anionic free radical)}$$

These free radicals produce tissue changes even under hypoxic conditions. When the oxygen concentration is higher, these radicals are diverted by reaction with oxygen.

$$RNO_2^{\cdot} + O_2 \quad RNO_2 + O_2^{\cdot} \text{ (superoxide)}$$

However, the superoxide radical is less reactive and tissues, including tumors, possess superoxide dismutase to decompose it.

6.3 EYE DISEASES

UV radiation produces harmful effects on skin, so it is no wonder that the eyes are also very sensitive to this type of exposure. Cataracts produced by ionizing radiation were found on survivors of Hiroshima and Nagasaki. These cataracts formed in a period of months to 3 years. Cataracts have also been found to be more common among patients with ankylosing spondylitis and tuberculosis that was treated with Ra^{224}. Generally, cataracts and some other eye diseases have been found to involve the action of free radicals [183].

Crystallin is the most light sensitive component of the eye. In the eye, crystallin functions as a lens, reducing the diffusion of light so it is focused clearly on the back of the eye. This property is due to its low refractive index, which is dependent on the composition and protein concentration of the structure. Following its formation, a cataract produces opacity of crystallin, resulting in a diffusion of light passing through the lens. The opacity is the result of the sulphydryl groups in the protein being oxidized. The yellow color of crystallin is due to its high concentration of tryptophan and its derivatives: kynurenines. Glucose complexed with 16-hydroxy-kynurenine absorbs light with a peak at 3,860 Å, which protects the retinal from UV exposure. At the same time it reduces chromatic aberration, allowing a clear image to be projected on the retina.

Crystallin is, therefore, a very sensitive component of the eyes that is routinely exposed to oxidative stress. Consequently, it should not be surprising that crystallin contains high levels of ascorbate, glutathione, and glutathione peroxidase to provide protection from hydrogenperoxide produced by UV-induced photosensitive reactions. Photosensitive reactions may also occur when atmospheric chemical pollutants are absorbed.

As a result of oxidative stress and aging, the malondialdehyde (MDA) content of the eye increases. As shown in table 6.6, Reactive oxygen species are involved in eye diseases. MDA accumulates while the antioxidant content decreases. Therefore, some treatments for cataract using purified superoxide dismutase or vitamin E should be effective at an early stage of cataract formation [106]. In experimentally induced cataract, the content of ascorbate decreases, but its administration in such cases increases oxidative stress [172].

Reactive oxygen species are also involved in other forms of eye diseases such as retrolenticular fibroplasia, ocular sclerosis, newborn retinopathies, chemical damage (naphthalene or acetophenone), and drug effects (chlorpromazine).

Table 6.6 The MDA content of eyes as a function of age and disease [172].

Age (years)	MDA content (nmoles per gram wet weight)			
	Control	Senility	Myopia	Diabetic cataract
40 - 50	0.75±0.09	2.33±.037	4.05±0.35*	6.15±0.63*
51 - 60	0.94±0.04	2.35±0.52	6.23±0.26*	8.33±0.75*
61 - 70	0.77±0.04	2.74±0.75	4.64±0.73*	7.76±0.45*
71 - 80	1.05±0.06	2.93±0.22	5.07±0.32*	7.30±0.70*

*Significantly increased over control.

Chapter 7

ANTIOXIDANTS

7.1 INTRODUCTION TO ANTIOXIDANTS

While there has been extensive research about the involvement of free radicals in the etiology of diseases, the evidence is clear only in a few cases. Most aspects of free radical involvement in disease, especially as causative factors are under intense debate.

Research on the biological role of free radicals and antioxidants done by the food production and processing industry, cosmetic manufactures, and veterinary medicine has attracted public attention. Many foods are marketed on the basis of their antioxidant qualities, either natural or enriched. Numerous books have been written on the healing role of nutrition. The scientific literature also contains a massive amount of information. In spite of all this popular and professional information that has been published, determining the practical aspects of antioxidants in nutrition and health is difficult [103].

Life on earth adapted to the increased amount of oxygen in the atmosphere through the selection of various antioxidant systems, both enzymatic and non-enzymatic. This list of antioxidants present in living organisms is growing through scientific study and synthetic antioxidants have also been discovered (figure 7.1). These antioxidants exhibit great structural diversity.

The second thing to note about antioxidants is their functional efficiency. Both enzymatic and non enzymatic antioxidants quickly scavenge free radicals and reactive oxygen species or prevent their formation. Free radicals are very reactive and have rate constants between 10^4 and 10^9 $M^{-1}s^{-1}$. Therefore, antioxidants must have equivalent rate constants. The antioxidant enzyme, catalase, has the most impressive rate constant. In fact the term catalysis was inspired by the action of catalase.

The third characteristic of antioxidants is their ability to function in cooperation. Effective protection from the action or formation of free radicals requires antioxidant activity in both aqueous and lipid environments, and in various parts of the cell structure. Antioxidants in the cell work in close harmony. When one antioxidant reacts with a free radical, another antioxidant is present to regenerate the first (figure 7.2B). In addition, these antioxidants may act by coupling to protect both the membrane and cytoplasm. Antioxidants in the cell possess high capacity and redundancy in order to always be available to protect the cell.

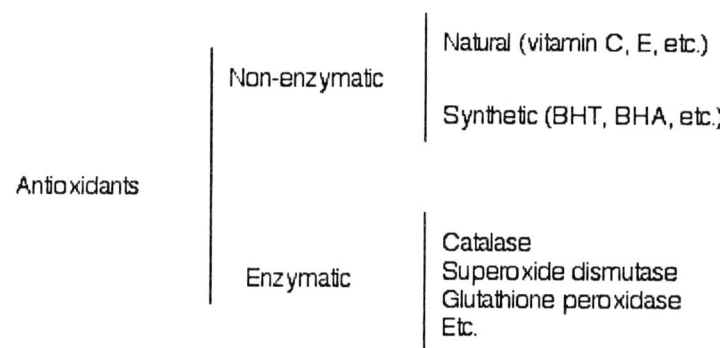

Figure 7.1. A classification of antioxidants.

A fourth characteristic has only become evident in the last few years. The role of antioxidants seems to be to sacrifice themselves to protect the essential molecules of the cell (DNA, rate limiting enzymes) or cellular structure (cell and organelle membranes).

The fifth characteristic of antioxidants is related to the sequential formation of reactive oxygen species. As shown in figure 7.2A, individual antioxidants are able to act at several points in the formation of organic peroxides.

The importance of antioxidant systems is emphasized by the fact that compounds with other primary biological roles (albumin, glucose, carnosine, taurine, uric acid, estrogens, carotenes, creatinine, vitamin A, dihydrolipoic acid, polyamines, fibrinogen) also can function as antioxidants. In addition to these endogenous compounds other natural compounds, such as flavonoids, phytic acid, and polyphenols have antioxidant capacity [30, 73]. This brief listing should emphasize the importance to the cell of protection from the actions of free radicals.

Not every antioxidant functions perfectly under all conditions. In fact, there is increasing concern that some antioxidants are actually prooxidants under certain conditions [63]. Cadenas [30] and Halliwell [73, 74] have recently shown that most of the nonenzymatic antioxidants are redox compound that function by the capture of an electron from the free radical. In rare circumstances, an antioxidant functions by being converted in the reaction to a live (but less reactive) free radical. This is clearly true for vitamins C and E and bilirubin, and is probably also the case for most redox compounds that react as pro- or antioxidants depending on their environment and concentration. Generally, the prooxidant property prevails for lower concentrations (10^{-6} to 10^{-5} μM) while the antioxidant property prevails for higher concentrations.

Antioxidants that become converted to a free radical are usually a part of a greater system that regenerates the original antioxidant. This is most clearly seen for vitamin E, which is regenerated by vitamin C and glutathione. Vitamin E accepts an electron from the radical, becoming a radical itself.

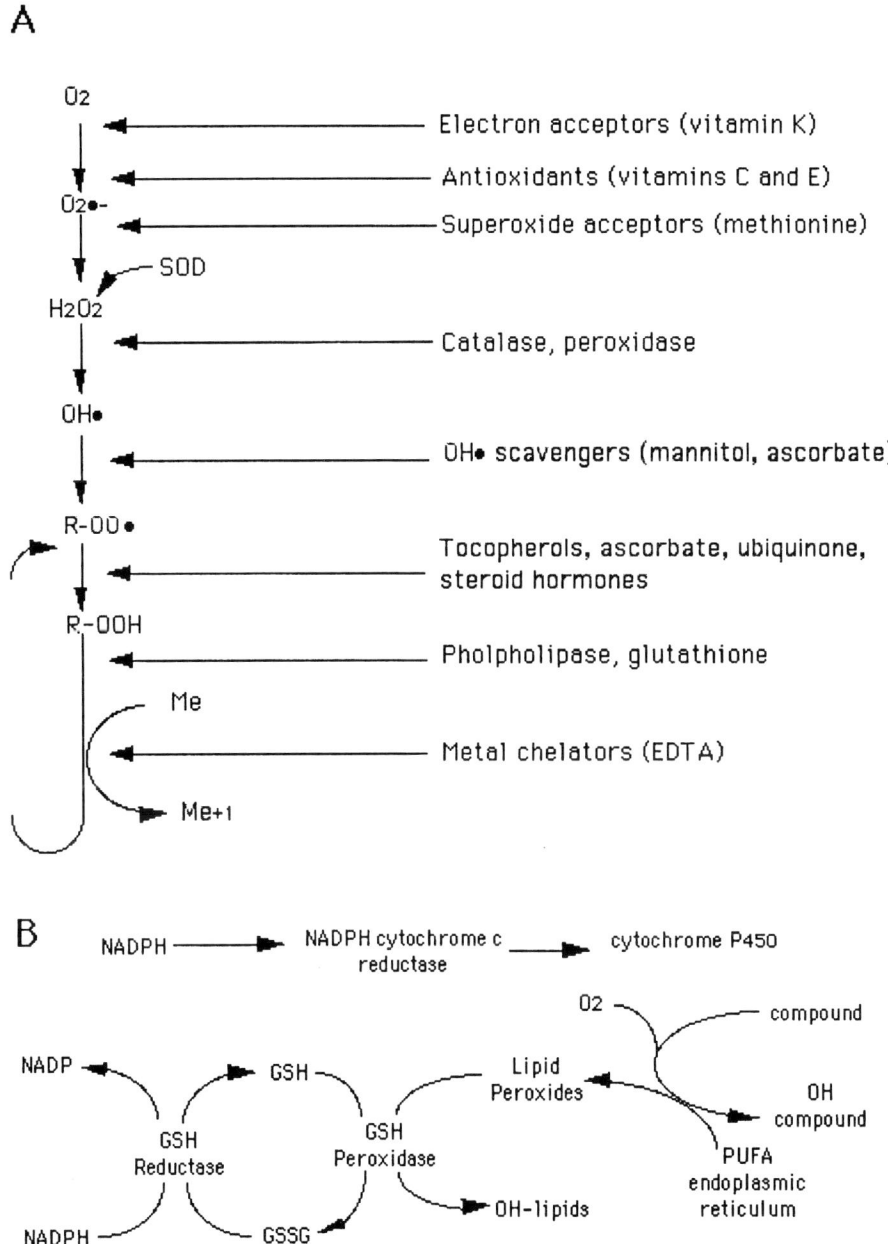

Figure 7.2. A, The evolution of reactive oxygen species and the antioxidants that may affect each step; B, Metabolic interactions among some antioxidants.

Vit. E + ROO• Vit. E• + ROOH

Vitamin E is equally able to react with singlet oxygen, superoxide, and hydroxyl. The scavenging action of vitamin E that occurs in tissue or cell suspensions do not result in the accumulation of a vitamin E free radical. Even in cell free reactions the vitamin E radical is eliminated by self reaction.

2 Vit. E˙   products

The resulting products are stable, harmless compounds or the original, active ones. Vitamin E radicals also readily react with other antioxidants (A) such as ascorbate or glutathione, which regenerates the vitamin E.

Vit. E˙ + A   Vit. E + A˙

Therefore, any redox compound that can regenerate a first line antioxidant (such as vitamin E or C) can also be considered an antioxidant. Vitamins E and C can be regenerated by many natural compounds, including uric acid, dehydrolipoic acid, and serotonin. This provides a powerful defensive strategy to keep the first line antioxidants available by using second line antioxidants to regenerate them.

However, this strategy may be limited by the physiologic concentrations of the antioxidants. For example, metallothioneins are tissue proteins that normally scavenge metallic ions but can also scavenge the hydroxyl radical at a high rate (10^{10} $M^{-1}s^{-1}$). However, metallothioneins are hampered in this reaction by their low physiologic concentration (10^{-7} M). On the other hand, albumin and glucose possess a lower capacity to scavenge hydroxyl, but their high concentration in the blood makes them effective antioxidants.

Other antioxidants function by capturing metallic ions that would otherwise act as prooxidants. Therefore, proteins like ceruloplasmin, albumin, lactoferrin, and transferrin are effective antioxidants because they effectively bind free metallic ions. Drugs such as desferroxamine and penicillin act similarly on iron and copper respectively [96, 135].

7.2 NATURAL ANTIOXIDANT ENZYMES

7.2.1 Superoxide Dismutase (SOD)

In the pathway from oxygen to organic peroxides (figure 7.2A) superoxide dismutase (SOD) is the first antioxidant enzyme encountered. SOD was discovered in 1968 and is one of the most unexpected and exciting events in the short history of free radical biochemistry. At Duke University, Fridovich and McCord were studying the reduction of cytochrome by xanthine oxidase. Up to this time, the existence of superoxide was only accepted by a small circle of nuclear physicists. However, the role of superoxide in biochemical processes was quickly accepted, as was the existence of an enzyme that quickly decomposed it: superoxide dismutase.

The discovery of SOD helped explain the existence of some peculiar intracellular proteins such as erythrocuprein in erythrocytes, hepatocuprein in hepatocytes, and cerebrocuprein in the brain. These proteins were originally thought to be a storage form of copper. Now it was clear they are stores of SOD. Many papers followed that demonstrated the existence of SOD in every aerobic cell on earth. It was also found that when a faculatative anaerobe was challenged with oxygen, SOD synthesis was quickly induced [19, 52, 121].

The first form of this antioxidant enzyme is believed to have evolved as early life had to find ways to defend itself from the increasing oxygen in the atmosphere. This is the manganese containing form of SOD. With diversification of life, new forms of SOD appeared that contain manganese (mitochondria), copper and zinc (cytoplasm), and iron (some bacteria) [121].

The discovery that superoxide is a physiologically occurring compound is a significant scientific advancement. This compound was difficult to detect because of its short half-life and the presence of SOD in biological systems to degrade it. SOD converts two superoxides to hydrogenperoxide.

$$O_2^{\bullet} + O_2^{\bullet} \xrightarrow{SOD} O_2 + H_2O_2$$

This reaction has a maximal rate of $K = 10^9 \, M^{-1}s^{-1}$. In the absence of SOD, this reaction proceeds very slowly ($K = 0.2 \, M^{-1}s^{-1}$). SOD is able to catalyze electron transfer between two molecule of superoxide due to the presence of copper, zinc, or manganese in the enzymes active center. X-ray crystallography shows SOD consists of two polypeptide subunits, each having a beta structure that covalently binds copper and zinc (table 7.1). Unlike other enzymes, SOD is not pH dependent. SOD has a high affinity for its substrate. Its specific *in vitro* inhibitor is diethyl dithiocarbonate ($I_{50} = 10$ mM). SOD is resistant to heat or denaturing compounds like urea or sodium dodecyl sulfate.

Table 7.1 Some Characteristics of superoxide dismutase

Type	Source	Rate constant ($M^{-1}s^{-1}$)	Molecular weight	Inhibitors
2 Cu-2 Zn	Cytoplasm	2.3×10^9	32,000	Cyanide, azide
4 Cu - Zn	Lungs	3×10^9	135,000	H_2O_2 in high conc.
2-4 Mn	Mitochondria	6×10^8	80,000	Sodium deoxycholate
Fe	Photobactria	5.5×10^8	40,000	H_2O_2

SOD is an inducible enzyme. Increasing concentrations of oxygen triggers its biosynthesis in bacteria like Streptococcus, Escherichia, or Saccharomyces and in mammals. Increased biosynthesis of SOD is observed in solid tumors as a defense against superoxide production by anthracycline cytostatics (adriamycin).

The oldest form of SOD, that containing manganese, is mostly located in mitochondria. It has long been observed that mitochondria from tumors contain little or no Mn-SOD. The rate of SOD biosynthesis in mitochondria is proportional to mitochondrial growth and respiration. Traces of iron may modulate mitochondrial Mn-SOD synthesis. But the presence of iron represses the induction of SOD in *E. Coli* challenged with paraquat, a herbicide that acts by producing superoxide. The regulatory action of iron ions on Mn-SOD biosynthesis is more complex. Chelators increase the synthesis of SOD 3 to 7 fold. These experiments suggest that, during evolution, organisms tried several mechanisms to handle the toxicity of free metallic ions. During these trials, several forms of metal containing proteins evolved with useful properties. While iron and copper ions promote the activation of oxygen, the proteins that contain them have the opposite effect: causing the decomposition of reactive oxygen species, allowing the safe transport of these metals, or acting as oxidases [127] (table 7.2).

Prooxidant metals are widely found in the environment. All of them are good catalysts for oxygen activation. However, when bound to proteins, they develop essential respiratory functions (hemoglobin, cytochrome oxidase) and antioxidant activity (SOD, catalase).

Table 7.2 Biological role of toxic metallic ions and their binding forms

Metal	Carrier	Required by:
Iron	Transferrin	Hemoglobin
	Albumin	Catalase
		Cytochrome oxidase
Copper	Ceruloplasmin	Superoxide dismutase
	Albumin	Hemocyanin
Cobalt	Albumin	Vitramin B_{12}
Selenium		Glutathione peroxidase

SOD has been found in significant amounts in all the cells of mammals. However, a decrease in SOD content has been found in some diseases, either due to reduced biosynthesis or inhibition (table 7.3). The decrease in SOD content is variable. Most (98%) of the SOD activity in mammals is found in the erythrocytes. These levels may remain normal while other tissues have a reduction in activity. For this reason, the determination of SOD activity of the blood is not a useful clinical test [121].

The experimental data concerning the protection of tissues against reactive oxygen species caused damage are impressive. So much so that about 10 years ago purified SOD was introduced as a therapeutic drug. Different pharmaceutical forms of the purified enzyme were produced with limited success as Orgatein (USA), Peroxynorm (Germany), and Epurox (Romania) or as nutritional supplements like Barley Green. Based on its biochemistry, SOD should be a powerful antiinflammatory. Therefore, SOD was administered for:

- Rheumatoid arthritis, osteoarthritis, Duchenne muscular dystrophy, the side effects of therapeutic irradiation, chronic cystitis, cataract, burns
- Autoimmune diseases like dermatomyositis, lupus erythematosus, Crohn's disease
- Paraquat intoxication
- Cerebral ischemia, Parkinson's disease, Alzheimer's disease

In the first group of diseases, good results were obtained, while in the other groups the results were variable. A major problem was its administration. SOD is a metabolizable enzyme having a plasma half-life of 2 hours. It does not penetrate tissue well and needs to be protected and transported to its target. This was attempted with encapsulation in liposomes or with polyethyleneglycol. J. R. Sorenson of Little Rock University claimed to have great success with clinical trials of a synthetic SOD. This compound is a copper complex with 3,5-diidopropylsalicylate (DIPS), which penetrates tissue well, and has antiinflammatory, radioprotective, and antitumor properties. Promising results were also had for the treatment of gastric ulcers and chemically induced diabetes. Unfortunately this drug proved to be very toxic, which hampers its further use. Some interesting effects that were noticed after the use of this synthetic SOD were the protection of lymphocytes against chromosomal scissions, the decrease in PMNL chemotaxis, and decreased gastrointestinal symptoms [182].

Table 7.3 Diseases in which decreased superoxide dismutase is found

Advanced senility	Bronchopulmonary dysplasia
Diabetes	Down syndrome
Cancer and leukemia	Alzheimer disease
Cerebrovascular modification	Parkinson disease
Rheumatoid arthritis	Lupus erythematous
Cataract	Osteoarthritis
Chronic cystitis	

Table 7.3 includes some rare congenital diseases. These diseases are usually fatal in the first decade of life. These diseases are autosomal recessives (Werner syndrome, Bloom syndrome, Down syndrome, Dublin-Johnson syndrome, ceroid lipofuscinosis, Franconi anemia, Ducheune dystrophy, cystic fibrosis, etc.). In these diseases, malformations, chromosomal aberrations, and an increased susceptibility to oxidative stress were demonstrated. For Bloom and Down syndromes, the gene on chromosome 21 that encodes Cu-Zn-SOD is over expressed. This results in the production of high amounts of hydrogenperoxide that may attack the DNA [115, 174]. The insertion of supplemental SOD genes into normal human fibroblasts produced mutations and increased peroxidation. A similar situation is produced by chronic intoxication with paraquat or anthracycline cytostatics. This is also seen in Alzheimer's disease. In these cases, a compensatory increase in other antioxidant enzymes (catalase and glutathione peroxidase) were observed in the blood [174].

7.2.2 Catalase

Catalase is present in all aerobic cells, especially in peroxisomes and mitochondria. Catalase degrades hydrogenperoxide to oxygen and water (1) or as a peroxidase (2).

$$H_2O_2 + H_2O \quad O_2 + 2 H_2O \qquad (K = 10^8 \text{ M}^{-1}\text{s}^{-1}) \qquad (1)$$

$$H_2O_2 + R\text{-}CH_2OH \quad 2 H_2O + R\text{-}CHO \qquad (2)$$

The large amount of catalase found in erythrocytes and liver are necessary because of the presence of metabolic processes that produce large amounts of hydrogenperoxide. The removal of hydrogenperoxide is strictly controlled in organisms. Three enzymes compete for it as a substrate: catalase, glutathione peroxidase, and leukocyte peroxidase. It is because of this that congenital deficiencies in catalase do not result in apparent harmful effects.

Congenital deficiencies in catalase become clinically detectable only under conditions of significant oxidative stress as may occur during an infection. Ulcers have been observed to develop in the mouth when infected with hydrogenperoxide producing bacteria. In this situation, there is insufficient catalase to provide protection and the hydrogenperoxide caused irritation that resulted in the formation of sores.

Like superoxide dismutase, catalase measurement is rarely used for clinical purposes.

7.2.3 Glutathione Peroxidase

Few studies focusing specifically on catalase have been published in the past 10 years. However, there have been many studies published on superoxide dismutase or glutathione peroxidase. Glutathione peroxidase catalyses the degradation of organic peroxides (ROOH), which may include hydrogenperoxide, or lipid peroxides (reaction 3, GSH = reduced glutathione, GSSG = oxidized glutathione, GSH Px = glutathione peroxidase).

$$ROOH + 2\ GSH \xrightarrow{GSH\ Px} R\text{-}OH + GSSG \tag{3}$$

The specificity of the enzyme for peroxides is low and the specificity for glutathione is high. Glutathione peroxidase competes with catalase for hydrogenperoxide. Glutathione peroxidase destroys hydrogenperoxide present at a low concentration (10^{-6} M) at physiologic glutathione concentrations (10^{-4} to 10^{-3} M). Catalase acts when hydrogenperoxide concentrations are high [18].

Glutathione peroxidase acts in close relationship with metabolic processes that are dependent on glycolosis. In fact, the formation of oxidized glutathione seems to be an index of oxidative stress at the cellular level. Oxidized glutathione also inhibits catalase [6, 123]. Therefore, oxidized glutathione needs to be removed. This is accomplished by coupling glutathione peroxidase to the reaction catalyzed by glutathione reductase, which regenerates reduced glutathione (GHS R = glutathione reductase).

$$GSSG + 2\ NADPH \xrightarrow{GSH\ R} 2\ GSH + 2\ NADP \tag{4}$$

Reduced NADPH is provided by the aerobic branch of glycolosis, the hexose monophosphate shunt (figure 7.2B).

The decomposition of hydrogenperoxide, the most common peroxide *in vivo*, is, therefore, a finely regulated competition between three enzymes that involves connection with metabolic processes, different activities based on different substrate concentrations, and different locations in the organism. In the cell, catalase is located in the mitochondria and peroxisomes along with glutathione peroxidase. In the cytoplasm and endoplasmic reticulum, glutathione peroxidase is found with superoxide dismutase.

There are two types of glutathione peroxidase, selenium dependent and selenium independent. Selenium dependent peroxidase is found int the cytosol and exhibits a high capacity to decompose hydrogenperoxide. Selenium independent peroxidase has a preference for organic peroxides. Another form of selenium dependent glutathione peroxidase acts efficiently on lipid peroxides.

The principle form of glutathione peroxidase contains four atoms of selenium, one in each of four subunits. The enzyme's molecular weight is 85,000. It decomposes hydrogenperoxide with a rate constant of 1.5×10^8 $M^{-1}s^{-1}$ and organic peroxides with a rate constant of 1 to 3×10^7 $M^{-1}s^{-1}$.

In addition to its role in protecting the mitochondria and membranes, glutathione peroxidase has a role in the biosynthesis of prostaglandins (fig. 7.3A)by regulating the production of the endoperoxides PGG_2 (Km = 12 µM) and PGH_2 (Km = 77 µM). Glutathione also regulates prostaglandin biosynthesis by catalyzing the formation of leukotrienes HPETE, HETE, and THETE (fig. 7.3B). Glutathione also regulates the formation of 15-HEPETE, a key intermediate in the biosynthesis of prostaglandins and thromboxanes (fig. 7.3C). The regulatory action of glutathione peroxidase for prostaglandin synthesis is so important that in selenium deficiency, lesions of the vascular endothelium appear and platelet aggregation is perturbed [120].

Unlike other antioxidant enzymes (superoxide dismutase and catalase), glutathione peroxidase is present in all tissues, although in variable amounts its connection to metabolic processes. The enzyme is inducible, as has been demonstrated in the lungs and erythrocytes of smokers [60, 90, 204]. Cigarette smoke contains high levels of free radicals, nitrogen oxides, and hydroperoxides, which act as promoters for glutathione peroxidase induction. Induction has also been noted for chronic oxidative stress (infection, exposure to chemical pollutants).

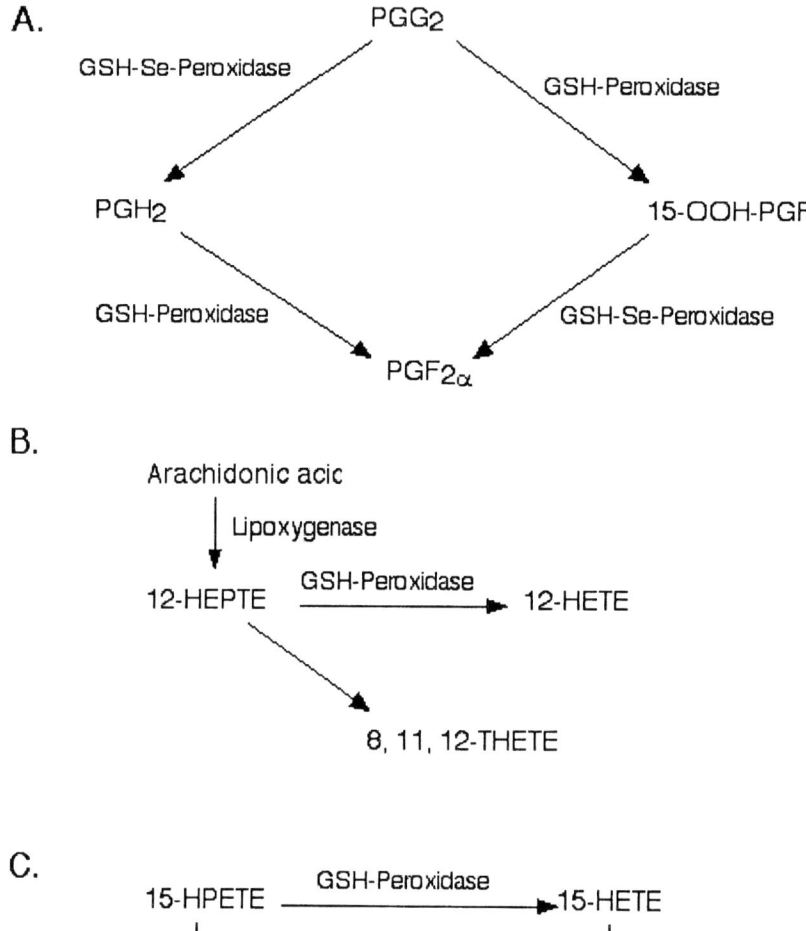

Figure 7.3. The influence of glutathione peroxidase on prostaglandin, thromboxane, and leukotriene formation.

7.2.4 Glutathione Transferase

Non-selenium dependent glutathione peroxidase also has a transferase activity. However, there are also a separate group of isoenzymes called glutathione transferase. Glutathione transferase has a molecular weight of 40,000 and has at least 5 isoenzymatic forms. Isoenzyme E is also known as epoxidase because it detoxifies organic epoxides. Glutathione transferase has a specificity for glutathione but a low preference for organic compounds. Glutathione transferase catalyses the conjugation of glutathione with a large number of organic, electophilic compounds, producing mercapturic acids. Known substrates for glutathione transferase include dinitrobenzene, epoxipropane, bromsulphalein, methyl iodide, bilirubin, and nearly 60 other compounds. This is a detoxification mechanism that occurs mostly in the liver. Glutathione transferase forms 5% to 10% of liver cytoplasmic proteins, and 2% of the soluble protein from kidney proximal tubules and of intestinal mucosa [12].

Aldehydes are a final product of the reactions of free radicals and peroxides. These are very reactive compounds that are decomposed by aldehyde oxidoreductase, another name for glutathione transferase (GSH = glutathione, GSH T = glutathione transferase).

$$GSH + R\text{-}CH=CH\text{-}COR' \xrightarrow{GSH\ T} R\text{-}CH(GS)\text{-}CH_2\text{-}COR' \qquad (5)$$

These α or β unsaturated aldehydes, like acrolein, are very reactive, producing inhibition of nucleic acid metabolism, protein biosynthesis, glycolysis, and mitochondrial respiration. They also possess antiviral, bactericidal, and antitumor activity.

Generally, mercapturic acids, the final products of glutathione transferase activity, are nontoxic. However, it has been discovered that the mercapturic acid formed using 1,2-dichlorethane (a solvent) or 1,2-dibromoethane (a mutagen) as the substrate are toxic. The 2-hydroxyethylglutathione conjugates attack nucleophilic compounds like DNA causing mutation and cancer.

The activity of glutathione transferase is strongly influenced by long-term treatment with barbiturates, paracetamol, phenacetin, and catecholamine derivatives (DOPA). Patients with diabetes, cholestasis, or chronic alcoholism have a reduced activity of glutathione transferase. The enzyme activity is increased in hyperthyroidism and decreased in hypothyroidism. For these patients, the administration of the above mentioned drugs must be done with care because of the body's sensitivity to the resulting mercapturic acids produced. Anesthetics, such as halothane, are metabolized to mercapturic acids that then modify the enzymatic activity. Quinones (vitamin K) and DOPA inhibit glutathione transferase.

Superoxide dismutase, catalase, and glutathione peroxidase are present at high levels in the erythrocytes and are nearly absent in the plasma. Their levels do not change very much during disease process except for those that affect their biosynthesis. Glutathione transferase is only released into the blood during lytic processes. This is a nonspecific process that occurs with infection, intoxication, and neoplasms. With the use of radioimmunoassay it was found that the physiologic level of glutathione transferase in the plasma is 4 ng/ml (3.3±2.4 U/L) and that this increases to 316 to 2,000 ng/ml in acute hepatitis, decreasing to about 50 ng/ml during remission. It remains at about 10±8 ng/ml in chronic hepatitis patients.

7.2.5 Ceruloplasmin

Ceruloplasmin is the single oxidase in the plasma and at the same time is the principle carrier of copper. This enzyme, a blue protein is abundant in plasma (approximately 300 μg/ml or 90 to 120 IU/L). It is an acute phase reactant protein, exhibiting a two to three fold increase during inflammation, infection, and neoplasm. Its physiologic function is not entirely known. Several activities of ceruloplasmin have been described including copper transport, oxidation of organic amines and catecholamines, oxidation of Fe^{2+}, and antioxidant activity against lipid peroxidation. *In vitro* experiments have shown that ceruloplasmin efficiently inhibits oxidation in various cell models and components (liposomes, phospholipids, polyunsaturated fatty acids, microsomal membranes). Recently it was shown that ceruloplasmin may oxidize LDL within the vascular endothelium. However, if it loses one copper atom (of the 8 bound per molecule) it completely suppresses LDL oxidation [40]. This variation in pro- and antioxidant activity may explain why ceruplasmin is sometimes considered a risk factor for cardiovascular disease.

The clinical measurement of ceruloplasmin is easily done and is used to diagnose Wilson's disease and for inflammatory condition or neoplasms.

7.2.6 Hemoxygenase

Hemoxygenase is an inducible enzyme that catalyses the removal of the heme ring of hemoglobin, producing biliverdin. This is a complex reaction requiring the coenzyme NADPH (Ho = hemoxygenase, Bil = biliverdin).

$$\text{Heme} + \text{NADPH} + O_2 \xrightarrow{\text{Ho}} \text{Bil} + \text{NADP} + Fe^{3+} + CO \tag{6}$$

R. Stocker proposed the hemoxygenase is an antioxidant enzyme because it is induced under oxidative stress conditions. To maintain homeostasis, cells respond to stress and metabolic perturbations by the production of inducible stress proteins. Among the stress proteins are heat shock proteins and superoxide dismutase. Stocker's hypothesis was based on the consequences of hemoxygenase induction: decrease in potential heme and heme proteins (hemoglobin, cytochrome P450) and increase in the level of bilirubin, which may have antioxidant properties. However, all these assumptions are controversial as a decrease in heme proteins is not beneficial for the organism (anemia, decreased detoxification capacity) and excessive bilirubin has harmful effects [184].

7.3 Non-Enzymatic Antioxidants

While there are only six enzymatic antioxidants (and not all of them are universally accepted) there are countless nonenzymatic antioxidants. The addition on nonenzymatic antioxidants greatly extends the organism's ability to defend itself against free radicals and reactive oxygen species. However, as we will see, some well known antioxidants can become prooxidant under the correct conditions.

7.3.1 Glutathione

Glutathione is a tripeptide (L-α-glutamyl-L-cysteinyl-glycine) that is one of the principle antioxidants in cells. It is also the carrier of free sulphydryl (SH) groups. In its reduced form, glutathione (GSH) is involved in multiple reactions and processes. This involvement is based on the great reactivity of the free sulphydryl group. Most other sulphydryl groups are associated with cysteine and are located in the active centers of enzymes. Unlike these enzymes, where the structure provides protection for the reactive group, glutathione has its sulphydryl group exposed to the environment where it is easily attacked by free radicals or reactive oxygen species. This makes glutathione the preferred target of free radicals or reactive oxygen species, resulting in its oxidation.

$$2 \text{ GSH} + 2 \text{ R}^{\bullet} \quad \text{GSSG} + 2 \text{ RH} \tag{7}$$

The metabolism of glutathione is illustrated in figure 7.4. This clearly shows the importance of this molecule as, in addition to its direct activity as an antioxidant, glutathione is a coenzyme for glutathione peroxidase, reductase, and transferase.

While glutathione was discovered in 1888, the molecule was largely ignored until the past 50 years. Gradually its multiple functions as antioxidant, cofactor, and substrate in detoxification became clear. All these functions make this a complicated molecule to understand. The plasma concentration of glutathione is low (about 85 µM) with 85% present in the reduced form [6]. The liver concentration is much higher (about 3 mM).

Synthesis of glutathione (fig. 7.4) requires γ-glutamyl-transpeptidase, which is a membrane bound enzyme that is widespread but especially abundant in the kidney. Cysteine oxidase is found to be another function of γ-glutamyl-transpeptidase.

Glutathione has a central role in chemical intoxication (acute or chronic) with compounds that produce free radicals (carbon tetrachloride, bromobenzene, acetaminophen). Many studies have demonstrated that the appearance of necrotic lesions and fatty liver infiltration occurs only after a severe drop in glutathione in the liver. Both of these are prevented by the administration of SH-containing substances such as cysteine, methionine, or cystamine. The protective role of free sulphydryl donors, including glutathione, has been demonstrated in isolated cells, tissue homogenates, animals, and humans. Administration of these compounds cause an increase in the tissue concentration of glutathione [186].

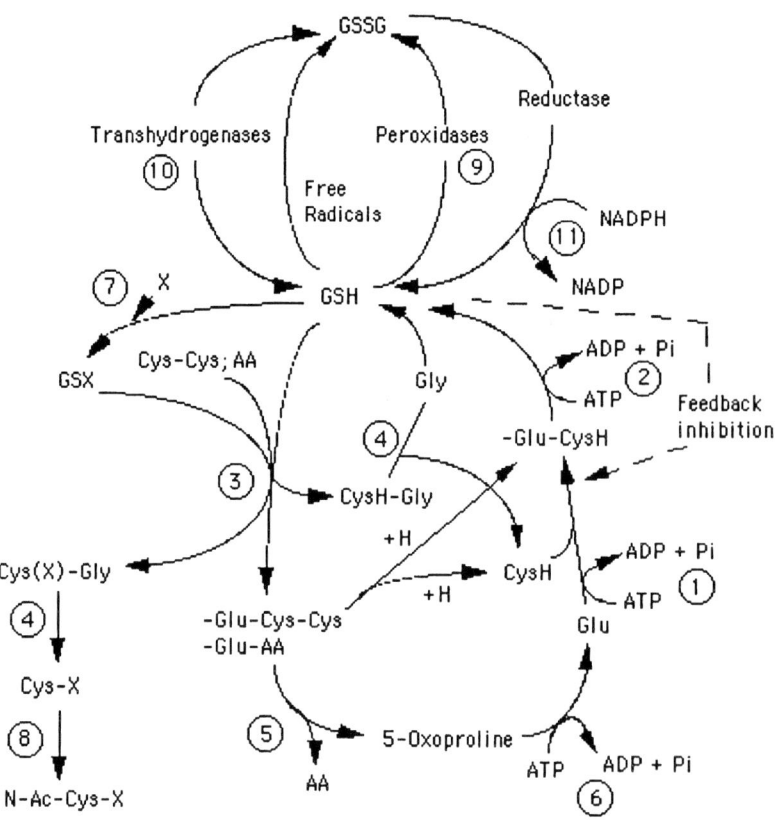

Figure 7.4. The biochemistry of glutathione (GSH). AA; amino acids. X; compounds that form conjugates with GSH. Numbers indicate enzymes: 1. g-glutamylcysteine synthetase, 2. GSH synthetase, 3. g-glutamyltraspeptidase, 4. dipeptidases, 5. g-glutamylcyclotransferase, 6. 5-oxoprolinase, 7. GSH S-transferase, 8. N-acetyltransferase, 9. GSH peroxidase, 10. GSH thiol transferase, 11. glutathione disulfide reductase.

As seen in table 7.4 there is a direct relationship between decreased glutathione and increased transaminase release, which is an index of cell lysis. The increase in erythrocyte lysis is strongly related to membrane peroxidation. Similar protection against chemical intoxication was demonstrated using N-acetylcysteine, but its clinical use is still under investigation [7].

The determination of glutathione in blood is not a common test. A simple method involves the determination of total free sulphydryl groups in plasma. This total is mostly due to free sulphydryl groups on albumin with a small contribution of glutathione and cysteine (100 µM). The physiologic limits of total free sulphydryl groups is between 380 and 500 µM). This parameter significantly decreases in patients with liver failure, in the final stages of cancer, or during toxic effects caused by treatment with cytostatic drugs [35, 48, 171].

The variation of free sulphydryl groups in plasma may be correlated with glutathione concentration in the liver, as was demonstrated by scientists from the national cancer research center in France following the administration of carcinogenic compounds to rats.

These compounds (4-dimethyl aminoazobenzene, methyl cholatrene) are metabolized with the intermediate formation of free radicals. During this metabolism in rats, a 100 fold increase in glutamyl transpeptidase activity was seen. This is similar to what is observed in a rapid growth process (neonatal or fetal). In the precancerous state, the concentration of glutathione in the liver increases 2 to 3 times in the first 48 hours after administration of the carcinogenic compound. The concentration of glutathione remains increased until the appearance of hyperplastic nodules and then, as the cancer develops, the glutathione level decreases. For differentiated tumors with a low growth rate, liver glutathione remains nearly normal. For undifferentiated tumors with a high growth rate, liver glutathione decreases to 50% of the initial values. It seems that a similar situation occurs in humans for whom increased early stage glutathione levels and decreased late stage glutathione levels have been observed [60, 126].

Table 7.4 Biochemical modification of plasma glutathione and transaminase in rats following the acute administration of chemical compounds [186].

Condition	GTP transaminase (U/ml)	Glutathione (mM/ml)	Erythrocyte lysis (%)
Saline (control)	10.2±1.3	5.5±0.8	12±6
CCl$_4$ (1 ml/kg)	31.8±2.7*	2.4±0.9*	26±9*
Paracetamol (500 mg/kg)	24.5±3.1*	3.6±1.3*	18±4
Cystamine (100 mg/kg)	14.6±1.5	6.4±0.7	14±5
Ccl4 + cystamine	21.7±1.8*	4.7±1.2	16±5
Parathion (2 mg/kg)	42.7±8.5*	2.6±0.7*	31±9*

As an effective antioxidant, glutathione plays an essential role in a variety of detoxification processes. This includes the amplification of peroxide damage that occurs with glutathione depletion, which increases the organism's susceptibility to cytotoxicity and affects drug interactions in neoplastic disease [6, 20, 48, 186].

Glutathione can also regenerate oxidized antioxidants, such as vitamins E and C (R$^{\bullet}$ = free radical, GSH = reduced glutathione, GSSG = oxidized glutathione).

$$R^{\bullet} + \text{vit. E} \quad \text{vit E}^{\bullet} + 2 \text{ GSH} \quad \text{Vit E} + \text{GSSG} \tag{8}$$

There is an inverse correlation between a low concentration of glutathione and a high level of peroxides. A decrease in glutathione with aging may be related to either the age-related increase in the oxidation rate (increasing the demand for glutathione) or to a decreased in overall glutathione turnover leading to reduced biosynthesis of glutathione. It is conceivable that the accumulation of toxic substances during senescence activate glutathione enzymes, resulting in intracellular glutathione depletion [41, 77, 199].

7.3.2 Vitamin E (α-tocopherol)

If glutathione is one of the principle nonenzymatic, water soluble antioxidants, then vitamin E is the main lipid soluble antioxidant. Its main role is to maintain membrane structural integrity. This role has become understood only in the last 10 years. Prior to this, vitamin E was regarded as a vitamin in search of a disease because deficiency was known in only a few animals and was not seen in humans. Even now, vitamin E deficiency is only seen under severe conditions such as severe malnourishment, especially in premature newborns [3, 49].

The controversy that remains regarding the role of vitamin E is due to the dependence on *in vitro* experiments using tissue homogenates. Only in rare, mostly congenital, diseases is a clear decrease in the blood level of vitamin E seen. Batten's disease (neuronal ceroid lipofuscinosis) is characterized by a progressive mental deficiency due to brain lesions and massive deposition of lipofuscin. In this disease, the vitamin E content of the blood is half normal without any clear causative mechanism [3, 170]. Hypersensitivity to hemolytic oxidative stress and exposure to ozone or hyperbaric oxygen has been observed under vitamin E deficiency [170].

Increased levels of vitamin E are found in adrenal glands, heart, testes, and liver. This distribution may result from its high lipid solubility. Intracellular vitamin E is associated with lipid membranes such as mitochondria and endoplasmic reticulum. These membranes also contain high amounts of ideal substrate for lipid peroxidation.

Like many antioxidants, vitamin E acts by sacrificing itself in a reaction with a free radical or reactive oxygen species. In the process it is transformed into a free radical (chromonoxyl). The presence of the tocopherol free radical, which has a long half-life, was demonstrated in liver with fatty infiltrations [170]. Vitamin E may be regenerated by other antioxidants, such as glutathione or vitamin C. Vitamin E is versatile as it is capable of reacting with free radicals, reactive oxygen species, or organic peroxyl radicals.

$$\text{vit E} + \text{ROO}^\bullet \quad \text{vit E}^\bullet + \text{ROOH} \tag{9}$$

The continuous regeneration of vitamin E by other antioxidants increases its benefit for lipid structures such as cell and organelle membranes and lipoproteins.

The most controversial aspect of the therapeutic use of vitamin E is the variable results obtained when it is applied to different diseases. For premature infants the administration of vitamin E (25 mg/kg im) significantly decreases jaundice and hemolysis and reduces lung complications (pulmonary dysplasia). In 1991, more than 1,000 papers appeared that claimed good clinical results for diseases such as duodenal ulcer, skin diseases, cirrhosis of the liver, and the preventive effects toward cancer and cardiovascular disease [3, 49]. Studies in rats and epidemiologic studies in humans have shown that vitamin E protects the lungs when exposed to ozone or cigarette smoke. The free radical content of expired air of smokers was reduced after they received an 800 IU daily supplement of vitamin E [94, 204].

In many experiments, induced vitamin E deficiency in animals has been found to affect the immune process (resistance to infection, antibody response, lymphocyte and neutrophile activation). The probable mechanism is a decrease in chemical signalling among immunocompetent cells because reactive oxygen species function as secondary messengers. If this were true it would be no wonder that vitamin E supplementation is claimed to produce beneficial results. However, it is not yet clear how much vitamin E is needed to achieve these results [49].

The human plasma concentration of vitamin E varies between 0.5 and 1 mg/dl (15 to 40 µM). Human studies on the effect of vitamin E are still inconclusive. A 1990 study was the first double blind, placebo controlled trial on the effect of vitamin E supplementation (800 IU per day) [13, 68, 81]. These studies supported earlier epidemiologic studies that indicated a lower incidence of infectious diseases in elderly people who had a high blood level of vitamin E. It has been difficult to accept that a single nutrient can be so important. This problem is so important that other studies were performed on larger populations (80,000 to 100,000). In all these studies, vitamin E showed strong protective effects, especially among people over 55 years of age. Those who consumed the highest level of vitamin E had the lowest risk for stroke, especially for older males with high LDL cholesterol. Studies have also been done to examine vitamin E's effect on cancer, cataract, Parkinson's disease, and arthritis. However, the results are generally inconclusive [3, 81, 100, 170].

The RDA for vitamin E was lowered from 30 IU to 15 IU in 1968, in part because, theoretically, it is difficult to get such an amount from the diet. However, many studies have shown that for an adequate defense against reactive oxygen species based only on this vitamin, a supplement of at least 20 IU per day is needed. Studies done in 1993 showed that a daily supplement of 100 IU per day reduced heart disease by 40%. There are communications that claim 800 to 1,000 IU of vitamin E should be used to boost the immune system of the elderly. The Council for Responsible Nutrition has recommended that people take several hundred IU per day. Adverse effects of high supplementation levels have been reported and include nausea, flatulence, diarrhea, headache, and weakness. The highest amounts of vitamin E are found in sunflower seeds (27 IU per 1/4 cup), hazelnuts (12 IU per 1/4 cup), corn, canola, safflower oils, (10 IU), and wheat germ (8 IU per 1/4 cup).

7.3.3 Vitamin C (Ascorbic Acid)

The therapeutic use of vitamin C has received extremely wide attention, especially among non-medical people. However, the evidence of the most efficacious amount of vitamin C that is needed is even less clear than it is for vitamin E. The labeling of vitamin C as an anti-infectious, detoxifying, antioxidant nutrient covers only a tenth of what is known about this vitamin [170].

The overall controversy about vitamin C began with discussions of its role as an antioxidant. As a reducing compound, vitamin C directly reacts with all reactive oxygen species ($O_2^{\cdot}$, $OH^{\cdot}$) and various lipid peroxides. In addition, it can restore the antioxidant properties of oxidized vitamin E or oxidized glutathione. This suggests a major function of the vitamin is to recycle oxidized antioxidants. If this is true, it helps explain the wide distribution in mammalian tissues. It is present in relatively high amounts in the adrenal and pituitary glands. Lesser amounts are found in the liver, spleen, pancreas, and brain [44, 166, 188].

When compared to other water-soluble antioxidants, vitamin C offers the most effective protection against plasma lipid peroxidation. As Ames and his coworkers have shown, during *in vitro* a challenge of plasma with a free radical generating system, ascorbate is the first antioxidant to be oxidized [4, 62]. When a free radical source appears in the blood plasma, ascorbate is the first line of defense. Ascorbate disappears first, before other antioxidants such as free sulphydryl groups, α-tocopherol, urate, etc. [62].

However, ascorbic acid can function equally well as either an anti- or prooxidant [44, 166]. Its actual role depends on its local concentration in the tissue and on the presence of metallic ions (especially iron or copper). *In vitro* experiments show that the induction of lipid peroxidation in the presence of ascorbic acid depends on its ability to reduce Fe^{3+} to Fe^{2+}, as ferrous ions are known to be potent free radical inducers. Oxidative modification of DNA bases are substantially enhanced in the presence of ascorbate.

At a physiological pH ascorbate exists as an anion. Its first step in reacting with reactive oxygen species is the formation of dehydroascorbic acid (through oxidation). It is not known if dehydroascorbic acid possesses both anti- and prooxidative properties. The formation of dehydroascorbic acid is advantageous for the antioxidant because it is easily regenerated to ascorbic acid. During oxidation of ascorbic acid, the ascorbyl free radical was detected using ESR [44].

The anti-infectious claims made for vitamin C are based on its ability to stimulate phagocytosis and transformation of lymphocytes, which should modulate cellular immunity. However, the action of vitamin C is complex. It seems that with nonactivated PMNLs, it stimulates the hexose monophosphate shunt. But with activated PMNLs, it decreases phagocytosis by inhibiting myeloperoxidase. The administration of 1 to 3 grams of vitamin C daily results in a significant increase in PMNL chemotaxis and lymphocyte sensitivity to mitogens is induced. Therefore, the effect of vitamin C on cellular immunity seems to depend on its plasma concentration. However, vitamin C in higher amounts can exert other effects, such as regeneration of other antioxidants, protection of lipoproteins from oxidation, inhibition of hyaluronic acid degeneration, and acceleration of collagen biosynthesis. Unfortunately, very high doses of vitamin C may also cause gastric irritation of the formation of gastric lesions.

The detoxification role of vitamin C is based on studies of the metabolism of xenobiotics by the hydroxylating system found mostly in the liver. This potential role is controversial. Vitamin C accelerates the detoxification of most xenobiotics, but by the same action, accelerates the activation of procarcinogens (benzopyrene) by cytochrome P_{450}. This accelerating effect of vitamin C on xenobiotic metabolism is only seen clearly in scorbutic animals. Thus it becomes something of a matter of semantics whether ascorbic acid promotes xenobiotic metabolism or if ascorbic acid deficiency inhibits xenobiotic metabolism.

The anti-tumor role of vitamin C is one of the most debated issues related to the vitamin. Dehydroascorbic acid (DHA) has cytotoxic effects on cells with at high metabolic rate, such as tumor cells. The observed radiosensitivity of lymphocytes seems to be due to the high amount of DNA in these cells. The administration of electrophilic radiosensitizers (metronidazol, misonidazol) is accompanied with a moderate dose of ascorbic acid to favor the production of cytotoxic free radicals.

This brings us to the most controversial aspect of dietary supplementation with vitamin C. In the 1960s Cameron and Pauling (who had already won two Nobel prizes) recommended the treatment of cancer with mega doses (up to 10 g daily) of vitamin C. Pauling claimed he had cured himself of colon cancer with this regimen. He eventually died of cancer, but not for another 20 years. Over 10 years ago the pros and cons of such a treatment were hotly debated and the medical community finally rejected the arguments in favor of it.

For more than 20 years, research into the possible role of ascorbate in preventing or curing cancer was avoided by any scientist who wanted to be taken seriously by the rest of the scientific community. However, in the late 1980s, this began to change. A number of epidemiological studies linked the frequency of various cancers with the vitamin C content of the diet [113, 188]. It seems that higher amounts of dietary vitamin C protect against throat, stomach, pancreatic, and breast cancers. Many studies have been done, but solid, undisputed results have not been found.

Since vitamin C may have a beneficial effect on detoxification and many cancers may be related to exposure to chemical pollutants, it seems logical that ascorbic acid would have some protective effect. This may explain the connection between low vitamin C content of the blood of smokers or in the blood of people exposed to mineral dust and the frequency of lung cancer [113, 142, 169].

Several studies have emphasized the protective effect of vitamin C [170, 188]. It is suggested that vitamin C may help prevent the formation of fatty plaques in the artery walls by inhibiting the oxidation of LDL. *In vitro* studies done by Dr. I. Jiald showed that vitamin C inhibits the uptake of LDL by macrophages 95% as well as does the drug Probucol [11, 95, 100]. Human studies have shown that 1 to 2 gram daily doses of vitamin C reduce the aggregation of platelets and their adhesion to artery walls. Other studies correlate higher blood levels of vitamin C with increased HDL and reduced serum cholesterol. It is suggested the reduction of cholesterol could be due to the shift of cholesterol metabolism towards the production of bile acids and their subsequent elimination from the body in the feces. Vitamin C also appears to reduce injury to the heart under ischemic conditions by inhibiting xanthene oxidase dependent reactive oxygen species formation during reperfusion. Incidents of deep vein thrombosis have been claimed to be reduced by ingestion of 1 gram of vitamin C daily [101, 166].

Numerous studies have also been done on the claims that large doses of vitamin C are capable of curing a cold and large numbers of people belief this is true. Vitamin C stimulates the production of leukocytes, which enhances the immunologic status of the elderly and stressed persons. It is observed that during infections or physical or emotional stress the vitamin C content of the blood decreases. Surgical patients have been observed to have about a 50% shorter recovery time when they are given 1 to 3 grams of vitamin C per day [13, 89, 97, 101].

In all of this study and arguments for and against the use of mega doses of vitamin C, the ideal amount humans need remains somewhat unclear. Vitamin C is widely produced in plants and in most animals. Humans are the exception as they lack an essential enzyme for the biosynthesis of vitamin C, so they must obtain it from the diet. This enzyme was lost due to a mutation millions of years ago, and humans have only survived because they eat fresh fruits, vegetables, and animals that are able to supply the needed vitamin. Anthropologists estimate that our ancestors obtained about 400 mg of vitamin C a day [44]. From the study of other animals, it is estimated that guinea pigs and primates eat the equivalent of 2 grams of vitamin C per day. Under conditions of stress, 7 to 10 grams per day are needed. Humans have a blood level of ascorbic acid between 0.7 and 2 mg/dl (30 to 90 µM). It is calculated that when under stress, animals (other than primates) synthesize the equivalent of 3 to 19 grams per day (normalized to a 154 pound human). Therefore, for a 154 pound person, a daily intake of 200 mg per day should maintain an adequate body pool.

The official recommended daily allowance (RDA), which is set to prevent deficiency disease is estimated to be as low as 60 mg for adult men and women. An intake of 100 mg has been recommended for smokers. There is also evidence that large portions of the United States population, especially the elderly and black males, obtain only 10% to 30% of the RDA [8, 94, 170].

J. Enstrom of the School of Public Health at the University of California, Los Angeles, published data from 10 years of research involving about 12,000 people [168, 206]. The men whose intake was about 300 mg of vitamin C per day had a 42% lower death rate from heart disease than did those who ate only 50 mg per day. No clear conclusions could be drawn for women [63].

Another question that must be asked is how much vitamin C can be taken without causing adverse side effects? No clear answer is known, for two reasons. First, as humans lack the enzyme needed to synthesize vitamin C, and because it is widely used in the body, no saturation limit for the vitamin has been found. Ascorbic acid is stored in the adrenals and in lesser amounts in the liver and brain. Any excess seems to be readily excreted. The second factor is individual variation. The circulating amount of vitamin C depends on several factors including intestinal absorption, oxidation and reduction in the blood, tissue uptake and release, and renal loss. Most cells and tissues take up vitamin C by active transport, resulting in intracellular levels that are often many fold higher than those in the plasma. Erythrocytes may carry as much as 30% of the total ascorbate in the blood. Many people can ingest 1 to 2 grams of vitamin C per day without any apparent problem. But there are others who experience gastric irritation and bleeding or a tendency to form kidney stones. It is suggested that a common sense approach is to maintain a daily intake of about 200 mg and, if you think it may be helpful, increase this to 1 gram if you have a cold or are under stress.

The best vegetable sources of vitamin C are cantaloupe and grapefruit (more than 100 mg per serving), strawberries, kiwi, mango, orange, and lemon (80 mg per serving), watermelon, broccoli, brussels sprouts, and cabbage (40 to 40 mg per serving).

7.3.4 Carotenoids and Vitamin A

The structurally related carotenoids and vitamin A enter the diet through plant sources and are strong antioxidants. Beta carotene is only one compound among a family of over 400 carotenoid pigments that give bright orange and yellow colors to fruits and vegetables. The structures of a number of the more common natural carotenoids are presented in figure 7.5. Their structure includes a polyisoprenoid chain that confers on the molecule an antioxidant property as a free radical scavenger. The carotenoids are lipid soluble, and are believed to work in conjunction with vitamin E as β-carotene is effective at low oxygen concentrations and vitamin E is effective at higher oxygen concentrations.

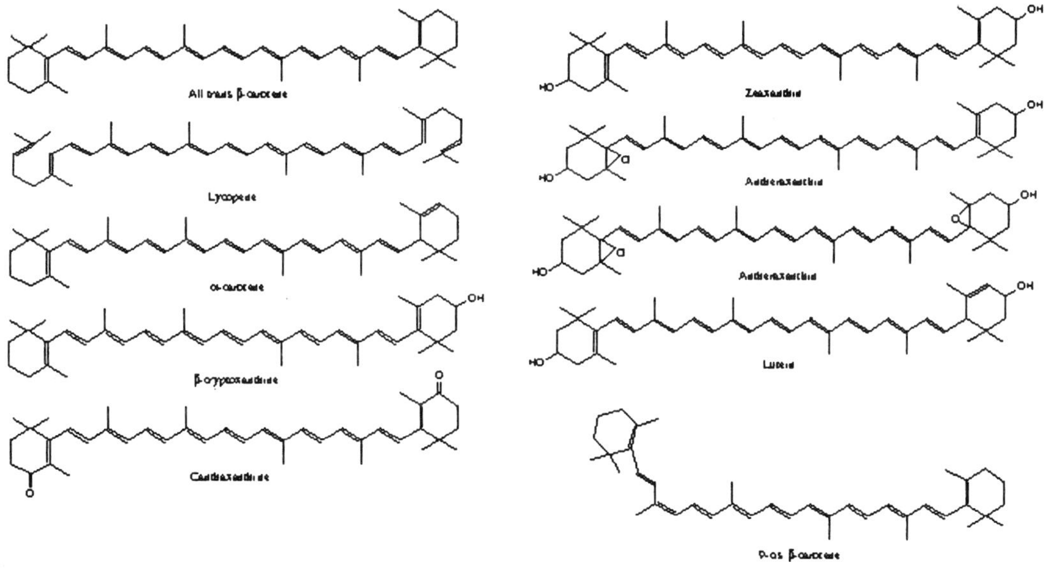

Figure 7.5. Some common natural carotenoids.

The efficiency of the quenching effect of β-carotene is high (1 mole quenched 1,000 molecules of superoxide) and the rate is fast ($k = 3 \times 10^{10}$ $M^{-1}s^{-1}$). This is made possible by the internal conversion of trans and cis bonds followed by a slow release of surplus energy. Superoxide initiates polyunsaturated fatty acid peroxidation with a rate of $k = 10^6$ $M^{-1}s^{-1}$. Therefore, the protection provided by β-carotene is due to its even faster rate of elimination of superoxide, even at a low superoxide concentration. In addition, this high antioxidant capacity does not require the regeneration of the molecule.

In *in vitro* studies, β-carotene has been suggested to have many positive effects [25, 107, 110, 169, 170].

- Provides protection against chemically induced cancers.
- Has antimitogenic effects against chemical and viral inducers.
- Inhibits precancerous proliferation of cells (phase G1)
- Inhibits lipid peroxidation in liver microsomes.
- Modifies lysosome permeability and, therefore, the release of acid phosphatase.
- Modifies cellular differentiation and cellular mitosis.
- Increases the killing capacity of T-lymphocytes towards tumors.

Observations on populations suggest carotenoids provide protection against some forms of cancer, especially lung cancer, some forms of heart disease, and eye diseases, especially cataract and macular degeneration [110, 183]. These properties make this compound an object of great interest, but also great controversy. Beta carotene is also the precursor (provitamin) for vitamin A. Beta carotene is converted into vitamin A in the body. Because of this, it is difficult to separate the activities of β-carotene and vitamin A when it operates in the body. However, Vitamin A seems to be a weak antioxidant that is far outperformed by β-carotene.

Without carotenoids, photosynthesis and life in an oxygen atmosphere would not be possible. Due to their conjugated double bond structure, delocalized π electrons can undergo a photon-induced transformation (excited singlet state) that allows them to transfer their energy to chlorophyll. Secondly the carotenoids provide photoprotection in photosynthetic structures. Carotenoids possess a highly unsaturated structure and are therefore able to extract or donate electrons to or from suitable substrates, leading to free radicals that in turn can react with oxygen and other compounds.

Alpha and β-carotene differ only in the position of a double bond in the B-ring (fig. 7.5). Other carotenes are hydroxylated on the aromatic rings. Due to their great light absorbance, carotenoids are intensely colored (yellow, orange, or red). Carotenoids can also accept excitation energy from singlet oxygen. The excitation state that results possesses a low energy and is unable to generate other free radicals. Consequently, the electronic excitation dissipates as heat. This property allows βcarotene to protect organisms against damage caused by the combination of light and oxygen. In fact, carotenoids provide an effective treatment for human patients suffering from erythropoietic porphyria, a condition in which free porphyrin accumulates in the skin. Free porphyrin is easily activated by light with the formation of free radicals.

All carotenoids rapidly react with oxidizing agents, but the specific reactivity depends on the length of the polyisoprenoid chain and the nature of the aromatic rings. The antioxidant property of carotenoids is easily demonstrated in *in vitro* experiments with organic solvents and lipophilic compounds. The *in vivo* situation is less clear. The concentration of carotenoids in mammalian tissues is generally much lower than those used to show antioxidant activity *in vitro*. In the physiologic situation, carotenoids are integrated into the lipid membranes. The presence of a carotenoid in a lipid membrane affects the membrane thickness, strength, and fluidity, as well as its barrier properties and permeability to oxygen. In addition, carotenoids may be stabilized by proteins as it has been found that carotenoids *in vivo* are more stable than are isolated carotenoids dissolved in organic solvents. All of this may help explain the difficulty of establishing without a doubt the antioxidant role of carotenoids *in vivo*. While this property is easily demonstrated in cell free systems, few papers have clearly show this effect in cell suspensions or whole animals [25, 110, 169].

Eating green leafy vegetables does not guarantee that the amount of vitamin A in the blood will increase. The release of carotenoids from food structures and their solubilization in the gut requires the presence of fats and bile acids. The absorption of carotenoids through the intestinal mucosa becomes a critical process. Once within the cells, carotenoids are metabolized by several routes to form retinal, retinoic acid and small amounts of breakdown products called β-apocarotenoids. However, only a fraction of the total carotenoids absorbed are converted. Carotenoids are taken up at different rates by different tissues, but little is known about the regulation of this process.

The beneficial role of β-carotene is attributed to its inhibition of polyunsaturated fatty acid oxidation [25, 169]. Such an indirect protective role against acute myocardial infarction was shown in a study involving 1,500 people from nine countries [102, 169, 206]. A similar role is believed to explain its protective effects against cancer. However, the *in vitro* studies on the anticarcinogenic role of carotenoids or vitamin A are not conclusive. High doses of vitamin A can reverse leukoplakia, but can also have toxic effects. Animal studies suggest that vitamin A can slow the metastatic spread of existing breast cancer in women with moderate dysplasia. Cells returned to normal in 43% of the patients using a vitamin A ointment. Retinal increases the effectiveness of cancer therapy using cytostatics or radiation by decreasing side effects (using 15,000 to 40,000 IU vitamin A and 10,000 to 20,000 IU of β-carotene) [23, 269]. Dietary retinoids are needed for repair of squamous lesions of the epithelium produced by chemical carcinogens. Experimental studies establish that retinoids inhibit growth of cervical cancer and cancer of the breast, lung, and pancreas in animals. Retinoids act as inducers of differentiation in established neoplasms. These data support the treatment of cancer with retinoids, but the subject remains controversial [23, 50, 110].

There is no RDA for β-carotene. However, several institutes, including the National Cancer Institute, recommend eating enough fruits and vegetables to provide 9,000 to 10,000 IU of β-carotene per day. The RDA for vitamin A has been established as 5,000 IU per day to prevent night blindness. People who are under significant oxidative stress (physical or emotional stress, smokers, history of cardiovascular disease, etc.) should take a higher dose. Some recommend 10,000 to 25,000 IU per day with half of this total taken as β-carotene.

Vitamin A is toxic. Taking more than 25,000 IU of vitamin A per day for several months may have side effects including loss of appetite, headache, blurred vision, hair loss, dry flaky skin, and possibly liver damage. In most situations these symptoms reverse themselves a few days after vitamin A intake is reduced. It seems that concurrent intake of vitamins A and E reduces the toxic effect of vitamin A. Specific receptors for retinoic acid are found on cells. These receptors are related to the steroid / thyroid hormone receptor superfamily. Therefore, it is possible that excessive amounts of vitamin A and especially retinoic acid may be teratogenic. A recent study found that lactating women in

Indonesia who took 300,000 IU should avoid large doses of the vitamin during the early stages of a subsequent pregnancy to avoid adverse effects on fetal development [50].

There are no reported toxic effects for β-carotenes. For some people, eating carotenes in amounts greater than 25,000 IU per day results in an orange coloration of the skin (carotenemia), which disappears without known toxic effects when the dose is reduced.

The best food sources for carotenes are 1 cup of carrot juice (25,000 IU); 1/2 cup of carrots or 1 pound of sweet potatoes (20,000 to 30,000 IU); 1 cup of mango, collard greens, cantaloupe, persimmon or 1/2 cup of cooked spinach or squash (3,000 to 8,000 IU); red pepper, vegetable juice, broccoli, apricots, watermelon (1,000 to 3,000 IU).

7.3.5 Melatonin

Melatonin is the latest fad on the antioxidant market. It was announced in 1995 as an "all natural miracle drug" and almost at once five books appeared to accompany this claim. It is claimed that melatonin improves sleep and slows aging.

Melatonin is synthesized in the pineal gland from tryptophan. Since its discovery 30 years ago, work on the hormone was restricted to a few laboratories and no clear function was established. It was generally stated that melatonin had some function in lipid metabolism and gonad function. The hormone shows a clear diurnal cycle with most of it being produced and secreted at night. This gave rise to its description as a chemical expression of darkness. The elevation of melatonin in the blood is roughly proportional to the duration of the daily dark period.

Light regulates pineal melatonin synthesis through the eyes. As light strikes the retina it activates a series of neurons that send signals from the eyes to the suprachiasmatic nuclei of the hypothalamus. During darkness the sympathetic neurons of the pineal gland release norepinephrine, which triggers melatonin synthesis. The melatonin level gradually increases during the first half of the night to reach a peak at midnight. It then decreases gradually and reaches daytime levels when light returns. We still do not know the physiological significance of the nocturnal pattern of melatonin biosynthesis.

In addition to light, extremely low frequency electric and magnetic fields or alterations in the static- or geomagnetic fields perturbs melatonin synthesis. Animal studies show that electromagnetic fields of 60 Hz can modify melatonin biosynthesis.

The antioxidant activity of melatonin is not high compared to glutathione or ascorbate, but increases significantly in their presence. Therefore, melatonin seems to enhance other antioxidants present in short supply [147]. As such, during the night melatonin should help clear out oxidative stress products produced during the day. This hypothesis may help explain the recuperative power of sleep. Nightly supplementation of melatonin may boost the performance of immune systems that are compromised by age, drugs, or stress. The melatonin level in the blood is higher in countries where breast cancer is rare. Chlorpromazine administration increases melatonin production and melatonin is decreased in some cardiovascular patients.

Melatonin dampens the release of estrogens and is able to shut down the reproductive system. There are great extremes in the observation, but melatonin may prolong the life span of cancer patients, probably by helping T-lymphocytes to proliferate and by emphasizing interleukin 2 action. A direct effect of melatonin causing sleep has not been clearly demonstrated.

7.3.6 Ubiquinone

Ubiquinone (coenzyme Q_{10}) is present in mitochondrial electron transport where it participates as an electron carrier in the lipid phase of the membrane. Ubiquinone is an essential electron carrier between other important compounds (flavoproteins and cytochromes). This function is based on the redox properties of ubiquinone, which also gives it an antioxidant property.

In the blood, ubiquinone is bound to lipoproteins, where it is mostly in a reduced form (ubiquinol or $CoQ_{10}H_2$). But in LDL, ubiquinol is easily oxidized to ubiquinone (CoQ_{10}). In fact, ubiquinol is the first antioxidant to be depleted when LDL is subjected to *in vitro* oxidative stress [30, 74].

Therefore, the antioxidant function of ubiquinone is important to protect LDL from lipid peroxidation, a process that is closely linked to atherogenesis and vascular disease. It seems that the $CoQ_{10}H_2/CoQ_{10}$ ratio should be a good measure of the oxidative stress present in plasma lipoproteins. This was the reasoning behind attempts to develop drugs based on ubiquinone to treat cardiovascular patients. However, no statistical difference has been found in this ratio when compared for smokers and nonsmokers. This also suggests the plasma lipoproteins are well protected against the oxidative stress of smoking. This may not be surprising given that the plasma and plasma lipoproteins contain antioxidant enzymes as well as vitamins C and E and many other compounds with antioxidant capacity discussed earlier.

7.3.7 Other Antioxidants

Uric acid and related purine compounds seem to have some antioxidant activity. These compounds are the end products of DNA and catabolism. The evidence that they have antioxidant activity is based on the work of Ames and coworkers [5]. In *in vitro* experiments they have shown that urate will scavenge hydroxyl radicals and protect polyunsaturated fatty acids, hemoglobin, and erythrocytes from oxidative stress. However, uric acid has a low solubility in plasma and any increase in its plasma level indicates the breakdown of DNA and increases the risk of the formation of urate stones in the bladder [171].

Albumin has astonishing antioxidant properties. Its major functions include the regulation of osmotic pressure and the transport of a range of substances, including iron, copper, fatty acids, bile acids, and some drugs. Albumin is a highly soluble protein with a molecular weight of 69,000 and a plasma concentration of about 40 mg/ml. Its half-life is about 20 days. The antioxidant property of albumin is expressed mostly by its ability to bind iron and copper ions that would otherwise stimulate oxygen activation or peroxidation. It is also able to scavenge hypochlorite ions and peroxyl radicals released by leukocytes. It scavenges singlet oxygen. Since the plasma volume is about 3 liters, the liver synthesizes about 3 grams of albumin daily. The albumin concentration in the plasma decreases in liver failure, when severe oxidative stress is taking place [126]. Albumin must be regarded as a sacrificial antioxidant because it is destroyed by reactive oxygen species when it functions as an antioxidant. Albumin is damaged by reactive oxygen species and the altered molecule is quickly replaced by *de novo* synthesis.

The albumin concentration in cerebrospinal fluid, aqueous humor, synovial fluid, and lung bronchoalveolar fluid is much lower compared to plasma. Consequently oxidative stress in these fluids is more dangerous as excess iron or copper is more likely to be able to initiate oxidation in inflammatory processes [71, 75].

An early event in tissue damage is increased vascular permeability. At a site of inflammation this is mediated by active compounds released by activated leukocytes. The synovial membrane shows increased permeability in inflamed joints. A benefit of increased vascular permeability is to increase the extracellular fluid content of plasma proteins like albumin, transferrin, and ceruloplasmin. Synovial fluid from inflamed joints contains significant amounts of these proteins. Allowing more albumin to cross a membrane barrier may help prevent excessive damage by reactive oxygen species.

Fibrinogen is a large molecule (molecular weight 340,000) and is involved in blood coagulation. Fibrinogen belongs to a family of proteins called acute phase proteins, such as ceruloplasmin, transferrin, and C-reactive protein. The antioxidant property of fibrinogen was reported after *in vitro* experiments found it scavenges superoxide, hydrogenperoxide, and hydroxyl radical. Fibrinogen also inhibits PMNL activation at concentrations close to its upper physiological limit (400 mg/ml). All acute phase proteins, including fibrinogen, increase in infections and in inflammatory conditions, where large amounts of reactive oxygen species are released. Fibrinogen then acts as an antioxidant by binding iron ions. Like albumin, it is damaged when it scavenges reactive oxygen substances. It is then easily degraded by thrombin or plasmin.

Lipoic acid is a derivative of octanoic acid and contains a disulfide bond. During its redox reaction it is converted to its reduced form (dihydrolipoic acid). Lipoic acid plays an essential role as a component of α-keto acid dehydrogenases, where its sulphydryl groups carry acyl groups. While lipoic acid is proven to be an effective antioxidant in *in vitro* systems, it also exhibits a prooxidant capacity. Lipoic acid has been shown to block tumor formation and phorbol ester activation of NF-KB, a known transcription activator of the HIV [108]. Under the name of thioctic acid, lipoic acid has been used in Germany for almost four decades for the treatment of diabetic polyneuropathy. The complications of diabetes may in part be related to oxidative stress triggered by the decompartmentalization of metal ions. Lipoic acid binds these ions and blocks metal ion simulated peroxidative reactions [169].

Metallothioneins are specific-metal binding proteins with a low molecular weight (6,000). Their polypeptide structures are characterized by a high content of cysteine residues. They do not contain aromatic amino acids. Metallothioneins are found in all living organisms. Mammalian metallothionein is a 61 to 68 amino acid peptide containing 20 cysteines, 7 lysines, and 7 to 10 serines. While human liver metallothionein contains zinc almost exclusively, that from kidney contains cadmium and copper. Metallothioneins have a long life span, as much as 15 years for cadmium-metallothionein.

There are three types of metallothioneins that are specific to plants, microorganisms, and mammals that are characterized by a high affinity for metals such as zinc, cadmium, and copper. Their concentration increases in liver or kidney following exposure to these metals and in some congenital diseases such as Wilson disease or Menke disease. Therefore, metallothioneins should be regarded as a cell mechanism to regulate the internal environment, relating mostly to detoxification. Since many of the metals bound by these proteins are strong prooxidants, these proteins should be considered as antioxidants [160].

Estrogens may act as antioxidants. Premenopausal women are protected against cardiovascular disease. Estrogens are active both in vascular and smooth muscle and the endothelium where they promote vasodilation by stimulating the synthesis of prostacyclin and nitric oxide. The low frequency of cardiovascular disease in middle aged women seems to be directly related to their diet. Many edible plant products, such as french beans, soy beans, and pomegranates contain phytoestrogens. *In vitro* experiments suggest these compounds may combat carcinogenesis. This may provide a partial explanation for the low breast cancer rate in Japanese women. The Japanese diet is high in tofu (a soy bean product), soy sauce, and miso. All of these are rich sources of phytoestrogens.

Among the reports of possible natural, endogenous antioxidants are carnosine and related compounds; phytic acid; amino acids like tryptophan, histidine, and taurine; ovothiols and ergothioneines; polyamines; estrogens; and α-hydroxy and keto acids. α-Hydroxy acids are widely used as active components in cosmetic products, especially anti-wrinkle creams. Adenosine, a nucleotide and component of DNA prevented post-traumatic epilepsy induced by the intracortical injection of iron ions with the subsequent production of reactive oxygen species. A clinical study confirmed early *in vitro* tests that showed antioxidant activity for adenosine. B vitamins, especially vitamin B_2 (riboflavin) also present significant antioxidant activity.

7.4 MINERAL ANTIOXIDANTS

Selenium has enjoyed great publicity in recent years for its antioxidant properties. Until its discovery as a cofactor for glutathione peroxidase, this element was regarded as solely a toxic material. Selenium toxicity in cattle has long been known. However malnutritive effects were also observed in animals living on selenium deficient soils. It has since been shown that selenium acts both independently and as a cofactor for glutathione peroxidase, with or without association with vitamin E. Selenium intensifies the biosynthesis of cytochrome P_{450} and dependent enzymes involved in the metabolism of xenobiotics in liver. Free selenium is present in the circulation as a conjugate with glutathione mixed disulfides [134].

The protection against selenium toxicity is central to its physiological role. Selenium toxicity is reduced in the range of 0.5 to 3.5 µg/kg by binding with glutathione or amino acids (especially methionine). At higher amounts, selenium becomes toxic. At lower amounts, selenium deficiency occurs. This is equally true for animals and humans.

In selenium deficiency, severe immunosuppression takes place along with the qualitative modification of neutrophils, T- and B-lymphocyte proliferation, and the production of antibodies and lymphokines [18, 120]. These modifications disappear as soon as selenium supplementation occurs. Therefore, the administration of selenium supplements of 250 to 400 µg per day improves immunologic performance by increasing chemotaxis and resistance to microbial and viral infections. In addition there are observations that selenium increases the rejection of grafted tumors and the cytotoxic effects of cancer cells toward natural killer cells [120]. Studies suggest a beneficial effect of cardiac muscle by increasing it movement and the biosynthesis of prostaglandin and thromboxane. Recall that selenium-dependent glutathione peroxidase is involved in the synthesis of prostaglandins.

The use of selenium as a therapeutic agent is hampered by its assimilation. While administration of 2 µg/kg of selenium methionine provided effective treatment in one experiment, the effect was equal to administration of 0.25 µg/kg sodium selenite. The LD_{50} for sodium selenite 1,700 µg per day while that for selenium enriched yeast is 5,000 µg per day. Therefore, selenium containing drug administration is usually done using enriched yeast. In Germany a synthetic form of selenium is sold as Ebselen.

Epidemiologic studies on selenium have mostly been done in Europe where an attempt has been made to correlate the administration of selenium, its blood level, and the frequency of diseases such as cancer and cardiovascular disease. An inverse relationship was observed, however, the conclusions are not clear [18, 120].

Important amounts of selenium are found in cereals (wheat and barley), bread, beef, pork, eggs, fish and seafood. In general, foods grown in coastal and glacial zones are lower in selenium [134].

Glutathione is one of the most important of the non-enzymatic antioxidants and is also part of an important enzymatic antioxidant system. Most of the benefits of selenium are associated with its being a cofactor in this enzymatic system. It seems clear that antioxidant activity is important in inflammation and detoxification reactions. It is less clear how selenium may have anticarcinogenic properties. This may be due to the following.

- Glutathione peroxidase may have a role through its influence on prostaglandin biosynthesis.
- It may stimulate detoxification of chemical carcinogens in the liver.

Dietary intake of selenium seems to inversely correlate with cancer risk. In South Dakota, a local with one of the highest soil selenium contents, there is a very low cancer incidence. In Ohio, where there is a very low soil selenium content, the cancer rate is nearly double that of South Dakota [120, 134]. The lower incidence of breast cancer in Japan may be partly explained by the high selenium intake because of the amount of seafood in the diet. Cancer patients tend to have low selenium levels and those with the lowest levels tend to have the most incidence of metastasis. Similar results have been observed for heart disease.

The recommended daily dose for selenium seems to be around 400 µg. Selenium toxicity in humans is rare. It seems to appear for doses higher than 5 mg per day. Symptoms of selenium toxicity include hair loss, garlic odor, fatigue, and thickened fingernails.

Zinc is a beneficial element that is required as a cofactor by more than 20 enzymes. Among these are superoxide dismutase and enzymes involved in nucleic acid synthesis. Zinc is also needed to maintain the structure of cell membranes [138].

Zinc seems to act mainly on the immune system. Cattle with zinc deficiency stop growing and are vulnerable to infections or die prematurely. The administration of zinc-enriched food is a remedy for some infections. Zinc deficiency in humans causes slow healing, impaired senses of taste and sight, loss of appetite, susceptibility for infections, and reduced fertility. Thus, zinc deficiency mostly affects organs involved in immune functions (spleen, thymus, and lymph nodes). It seems that zinc is especially depleted during upper respiratory infections that are accompanied by fever. Its anti-inflammatory actions are related to the inhibition of histamine release from immunocompetent cells. A relieving effect was also reported for rheumatoid arthritis. Clear improvement in the susceptibility to infections has been reported for persons over 70 years of age who received a 220 mg zinc supplement per day [42, 73]. Healing was improved by a zinc supplemented diet (150 mg per day) with a shorter recovery time. Similar results have been reported for ulcer patients [138].

Zinc's improvement of immunity is claimed to also provide resistance to cancer. Cancer patients often have low zinc levels (and are higher in copper). Additional zinc added to animal diets reduced the risk of cancer when the animals were exposed to chemical carcinogens [138]. Zinc enriched animal diets also protected against the toxicity of carbon tetrachloride and metals such as lead and cadmium.

Zinc is found in high concentrations in testes, semen, and the prostate. Zinc deficiency affects sperm count and motility. Significant improvements in male fertility have been seen with a 50 mg per day zinc supplement [52].

Zinc doses of 2,000 mg per day result in gastrointestinal distress symptoms such as nausea and vomiting. However, true toxicity does not occur unless the zinc is inhaled or even larger amounts are ingested. Large amounts of excess zinc are stored in tissues and bound to metallothionein [138].

Like selenium, the zinc content of food depends on the soil content. Therefore, mild zinc deficiency, even in the USA, may be common. Zinc absorption is hampered by the intake of alcohol, steroids, and other drugs. The RDA for zinc is 15 to 20 mg per day, but the Council for Responsible Nutrition recommends 50 mg per day. High doses of zinc may interfere with the absorption of copper and iron.

Copper circulates in the blood bound to ceruloplasmin and albumin. Copper is needed as a cofactor for superoxide dismutase and for biosynthetic processes of collagen, catecholamines and melaninic pigments [40]. Copper has been implicated in inflammation and arthritis. Copper bracelets have been a common folk remedy for rheumatic disease. Those who practice it believe minute amounts of copper are absorbed through the skin.

Copper seems to be a first line of defense against reactive oxygen species released in inflammatory conditions. Superoxide dismutase and ceruloplasmin are both effective antioxidant enzymes that require copper.

Some organic complexes of copper exert antiinflammatory and analgesic actions. These complexes activate the opiate receptor and cross the blood-brain barrier. Other copper complexes imitate superoxide dismutase, and exert antidiabetic, anticonvulsive effects. Some of these are used as drugs such as Cupralene and Cuproxalene [182].

Regardless of the causes of infection or inflammation, plasma copper increases beginning at the onset of the condition. It remains at a high level until cure or recovery. These observations require more research to understand their significance [182].

In spite of the beneficial role of copper, copper toxicity is reported in hereditary Wilson disease, in which metal depots appear in various parts of the body, including the eyes. Accidental ingestion of large amounts of copper (usually from acidic beverages or hard liquor distilled in copper pipes) results in intestinal irritation and nausea, vomiting, headache, and dizziness. Because the copper absorption mechanism becomes saturated, much of the metal remains unabsorbed and is excreted in the bile.

Copper deficiency is rare and produces anemia, impaired immunity and bone disease. There is no RDA for copper, but an adequate daily intake for adults averages 2 to 3 mg. The Council for Responsible Nutrition recommends 5 to 10 mg daily. Good sources of copper are liver, shellfish, meats, nuts, and whole grains. Cooking utensils and copper water pipes may contribute additional dietary copper.

Manganese is an essential mineral. It is a cofactor for superoxide dismutase and is necessary for animal reproduction, glucose tolerance, bone formation and catecholamine synthesis. Manganese also influences immunity and is found to be low in women with osteoporosis. The Council for Responsible Nutrition recommends and intake of about 10 mg per day. Manganese is found in significant amounts in fruits, seeds and whole grains, milk, and organ meats.

7.5 PLANT ANTIOXIDANTS

Compounds with antioxidative properties are widespread among plants. The rich content of antioxidants in plants may be the basis of the historical therapeutic use of fresh fruits, roots, and leaves (as teas or extracts). The use of plants (herbals) in therapy and alternative medicine remains widespread in popular culture. Many traditional herbal remedies have become the basis of modern pharmaceuticals.

Flavonoids are the best known plant antioxidants. The therapeutic use of flavonoids is accepted and several drugs contain high amounts: Troxevasin, Glyvenol, Rutozide, Benuruton (all of which are used to treat varicose and fragile veins). Flavonoids are a heterogenous group of compounds that often have contradictory therapeutic properties.

- Most are effective antioxidants against all types of reactive oxygen species.
- They protect and potentiate ascorbic acid function.
- They protect collagen against oxidative degradation.
- They inhibit cyclic nucleotide-dependent phosphodiesterase.
- They inhibit the activation of carcinogenic hydrocarbons.
- They inhibit PMNL phagocytosis and exert antiiflammatory properties.
- They possess antiadhesive and antiplatelet aggregation properties.
- They stimulate prostaglandin synthesis and epoxide hydrolase.
- They possess antilipemic properties and decrease cholesterol levels in the blood.
- They stimulate ammonia detoxification in uremic patients.

Flavonoids include over 4,000 compounds that include chalcones, flavonols, anthocyanidins, isoflavonoids, and their derivatives. In plants they function as visual signals (color) or as a defense against being eaten. Their antioxidant properties were discovered in the early 1930s when they were called vitamin P [8, 30].

Flavonoids moderately scavenge reactive oxygen species. These compounds have a high affinity for iron, so their antioxidant activity is mostly due to their chelating effects. Flavonoids differ in their antioxidant capability: rutin > hesperetin > quercitin > naringenin. The capacity of flavonoids to modify membrane-dependent processes (peroxidative damage) is related not only to their structure but also to their ability to interact with and penetrate the lipid bilayer.

Albert Szent Gyorgy, who discovered vitamin C, observed that in plants flavonoids occur in close relationship with ascorbic acid. Flavonoids and vitamin C function in a reciprocal protective system that enhances their antioxidant activity. Flavonoids also protect vitamin C against oxidation by binding iron or copper ions.

Flavonoids have long been known to be circulation enhancers and capillary protectors [37, 44, 166]. Flavonoids are also useful in the treatment of inflammatory conditions by scavenging excess reactive oxygen species and protecting cells from oxidative attack. Therefore, flavonoids are useful in the treatment of asthma and allergies connected with inflammation. The use of flavonoids in some, but not all, cases of cardiovascular disease or cancer seems to be beneficial [152].

The use of flavonoids to treat inflammations caused by infections is supported by their antiviral and antibacterial activity, which is based on their ability to strongly bind proteins. Flavonoids *in vitro* inactivate viruses (herpes, polio, rhinovirus, influenza) [1, 74, 107].

Quericitin, rutin, and tanins are the most studied flavonoids, but others are equally effective. Flavonoids are found in a wide variety of plants including cranberries (and other berries), citrus fruits (quercitin and hesperidin), buckwheat [4, 74, 152]. Green tea is especially rich in tannins (catechins) that are useful for relieving rheumatic fever.

Green tea is the most useful and least costly way to introduce antioxidants into the diet. The usefulness of exotic antioxidants such as ginko biloba, atragallus, pycnogenol, ginseng, and other plants is mostly due to their flavonoid content.

Phenol and polyphenols are another class of natural products with an aromatic structure that is close to that of flavonoids. Unlike flavonoids, some phenols are easily oxidized and stimulate oxidation of peroxidation. Polyphenols are also of biochemical interest because of their demonstrated antimicrobial, anti-tumor, and free radical scavenging properties [8, 37, 107, 184].

7.6 SYNTHETIC ANTIOXIDANTS

With the discovery in the 1940s and 1950s of the peroxidative reactions affecting food, rubber, and other manufactured products, attention was turned to finding practical solutions to prevent this [123, 140, 192]. The cost of natural antioxidants, such as vitamins C and E are high, so industry turned to the development of synthetic antioxidants.

The most important synthetic antioxidants used in food are BHT, BHA, and propyl gallate. Their use has allowed the long-term storage of food without lipid peroxidation (rancidity). These synthetic antioxidants are metabolized in the liver, and, therefore, do not provide protection *in vivo*.

Chapter 8

PHYSIOLOGIC EFFECTS OF OXYGEN STRESS

8.1 INTRODUCTION

Previous chapters have discussed the formation and preventive mechanisms for the physiologic production of reactive oxygen species. While reactive oxygen species are continuously being produced in the organism, a healthy individual generally possesses a mechanism in sufficient quality and quantity to prevent them from causing harm. Never-the-less there is evidence that even among healthy persons, if they are subjected to sufficient oxidative stress, harm can be done. In most of these cases, the defenses are only slightly exceeded so that the damage is minor and quickly repaired when the stressor is removed. In this situation, no clinical evidence of disease becomes apparent. If the stress is larger or becomes chronic, damage may accumulate and clinical consequences result.

8.2 PREGNANCY AND THE NEONATE

Pregnancy is a stressful condition for women involving a wide array of biochemical and biological modifications, including oxidative stress. However, most women are biologically able to adapt to the situation and have enough antioxidant capacity to keep the oxidative stress effects within a safe, reversible range. In table 8.1 are listed the levels of lipid and lipid peroxidation in pregnant women and newborns. The table clearly shows that for both pregnant women and newborns, the peroxide levels are increased. For new borns increased levels of bilirubin and peroxides can reach dangerous levels as severe irreversible lesions develop in the brain (encephalopathies). In both cases, the highest levels of peroxides were found in the severe conditions. Supplemental antioxidants should be considered as part of the treatment in these cases.

While toxic pregnancy involves a breakdown of the antioxidant defense, the increase in peroxidation in new borns is due to the fact that the antioxidant defense is not completely active in new borns. The liver antioxidant systems are the main protective weapon for the adult, but the newborn's liver does have the capacity to deal with the reactive oxygen species generated as excess erythrocytes are destroyed. The brain is most affected because, even in the adult, this organ has a low antioxidant capacity.

Table 8.1 Lipid and lipid peroxide in the plasma of pregnant women and new borns. After Yagi [203] and Olinescu [123].

Pregnancy	Peroxides (nmoles MDA[a])	Total lipid (mg/dl)
Non pregnant	3.74±0.13	550±35
1st trimester	3.12±0.12	527±27
2nd trimester	4.15±0.21	764±42*
3rd trimester	4.82±0.32*	963±47**
Toxic pregnancy	5.38±0.52*	863±37**
New borns		Bilirubin (mg/dl)
1st day, mild jaundice	10.03±2.06	7.2±1.5
1st day, jaundice	21.23±1.89**	10.09±0.61*
2nd day, hyperbilirubinemia	24.05±4.3**	17.10±1.20**

[a]Malondialdehyde
*$P<0.05$
**$P<0.01$

8.3 PHYSICAL AND EMOTIONAL STRESS

Oxidative stress occurring in athletes under intense training or in competition is a long disputed matter. Beginning in the 1980s, American scientists began to study oxidative stress in both athletes and animals. It had long been known that physiological, biochemical, and biological modifications occur during physical exertion. The formation of lactic acid is one such consequence. Others include cellular lysis with the increased elimination of proteins, hemoglobin, carbohydrates, etc. In the urine. These modifications are mainly the consequence of a mild, reversible increased renal permeability combined with increased cell lysis. The greater the effort and the lower the degree of conditioning leads to greater lysis and excretion. However, even in trained athletes, these modifications are detectable. These modifications also include an increased muscular sensitivity to pro-oxidative agents (infections, inflammations, stress).

In athletes involved in severe physical effort, the main source of reactive oxygen species is due to an increased mitochondrial respiration. This has been detected as increased plasma lipid peroxidation, pentane in exhaled air, and through ESR [55, 129, 167]. A hypothetical sequence of events leading to the formation reactive oxygen species and involving xanthene oxidase during exercise is presented in figure 8.1. Since reactive oxygen species are implicated in all inflammatory conditions, there may be a link between the physical exertion of athletes and professional dancers and a later tendency to develop rheumatic and other inflammatory diseases. It has been demonstrated that physical exertion combined with aging both increase sensitivity to oxidative stress and lead to the accumulation of lipofuscin pigments. This process is strongly influenced by antioxidant intake and caloric restriction, which seems to increase antioxidant efficiency in tissues [120, 150, 205].

Physical effort always involves a certain amount of emotional stress. The response to emotional stress varies strongly with the type and duration of the stress and with the individual. The activation of homeostasis mechanisms to neutralize modifications due to oxidative stress also varies among individuals. Therefore, during experimentally induced painful emotional stress, biochemical modifications occur including cell lysis (with the release of intracellular enzymes such as transaminase and cathepsins) and the increased formation of peroxides. As shown in table 8.2, these processes are all influenced by the administration of antioxidants [111, 114, 129]. The results shown here clearly show that painful emotional stress produces peroxidation of cell membranes (peroxides and Schiff bases, which is a measure of lipfuscin) and cell lysis (transaminase and cathepsins). The continuation of this stress condition lead to morphologic lesions in the heart and blood vessels. However, these were still reversible with the administration of antioxidants.

Intense emotional stress is also accompanied by the release of catecholamines (adrenalin), which, when present in large amounts, can be oxidatively metabolized, producing free radical intermediates [21]. The relationship between free radical formation and emotional stress is not understood. But it is known that glutamic acid, which has neurotoxic properties, may be involved. This amino acid is involved in aerobic glycolysis and the tricarboxylic acid cycle. Glutamic acid is transformed into gamma-aminobutyric acid (GABA) and later into oxybutyric acid (GOBA). Under extreme emotional painful stress, the conversion of glutamic acid to GABA and GOBA is increased. The experimental administration of 100 mg/kg GOBA to animals undergoing painful emotional stress protected against peroxidation. However, glutamate itself is a prooxidant. In addition, this amino acid exhibits neuroexcitatory properties that have only recently been discovered and are not clearly understood [45, 118, 174].

8.4 AGING

Aging has been widely studied and many new scientific papers and popular books on the subject are published every year. Interest in this subject is enhanced by the fact that in the United States, the older adult portion of the population is rapidly increasing. At the International Congress of Biochemistry held in 1989 in Prague, Professor B. L. Strehler characterized aging by the following features:

Table 8.2 Biochemical modifications in the heart and plasma of rats exposed to painful emotional stress (PES) [111].

Condition	Cathepsin(μg/mg/h)	Peroxides (nmoles/mg)	Schiff base (units)	Plasma transaminase (μmoles/ml)
Control	0.74±0.05	16.7±1.4	10.5±3.5	2.40±0.29
PES	1.82±0.06	47.6±2.4*	27.6±4.2*	6.10±0.12*
PES + 10 mg vitamin E	0.80±0.03	28.7±2.5	13.2±1.8	2.90±0.16
PES + 2 mg indomethacin	0.83±0.06	32.8±2.9*	15.3±2.9	3.17±0.12
PES + 50 mg Na selenite	1.38±0.03	31.8±3.5*	20.7±1.6*	4.18±0.20*

*Significantly increased over control

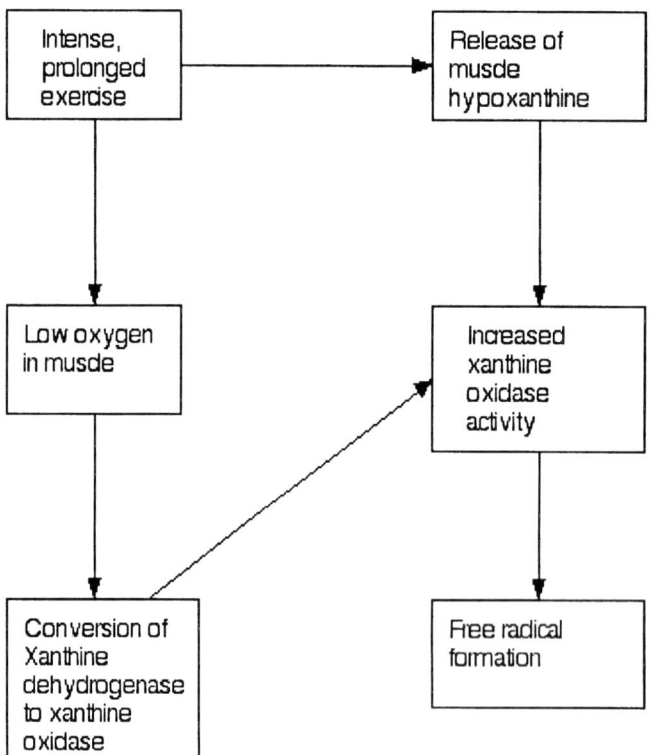

Figure 8.1. Probable role of exercise in the production of free radicals.

- Aging is universal. It occurs in all living organisms in spite of some speculations regarding eukaryotes.
- There is some degree of genetic programming for aging. Human fibroblasts in culture exhibit a maximum of 50 divisions. Genetic programming relating to aging may affect the metabolism of the individual [25].
- Aging is clearly seen in cells that do not divide, such as neurons and some muscle cells. At a certain age the accumulation of lipofuscin in these cells becomes evident. These brown, insoluble pigments are very complex. Their composition includes cellular detritus, lipid peroxides, and lipid peroxide decomposition products (aldehydes, Schiff bases). Lipofuscin pigments accumulate linearly with age, especially in the brain and heart. At 80 years of age these pigments reach 70% of the cytoplasmic volume of neurons and 6% of the myocardium. These pigments accumulate more rapidly in animals with short life-spans or a low content of vitamin E.
- During aging a decrease of collagen biosynthesis takes place.

The rate of the mitochondrial formation of superoxide and hydrogenperoxide for different mammalian species is inversely correlated with the longevity of the species. The steady state of mitochondrial DNA oxidative change is about 16-fold greater than the rate for nuclear DNA. Oxidative damage to nuclear DNA increases in proportion to species-specific basal metabolic rates. It is likely that decreases in oxygen consumption by humans due to dietary restriction are associated with lower rates of nuclear DNA oxidative damage [5, 34, 88, 181].

Erythrocytes have a mean life of 120 days. As they approach this age, both qualitative and quantitative changes occur, such as an in increase lipid peroxidation and cytosolic viscosity, and decreased hemoglobin content. There is an overall increase in sensitivity to the effect of oxidants. At the same time the intracellular content of antioxidants decreases [203].

Based on this, Professor Denham Harman developed the hypothesis of increasing free radicals with aging, concomitant with a decrease in antioxidant protection. This hypothesis was formulated in 1956, but remains popular. Data showing a connection between lipid peroxidation and aging events continue to accumulate. Peroxidation has been clearly implicated in cataract formation (see table 6.6). Patients with 21 trisomy age rapidly and their brain is affected with Alzheimer-like lesions. The determination of superoxide dismutase content in these persons gives conflicting results. Some show very greatly decreased values and others show twice the normal amounts as the cells attempt to fight the formation of superoxide and other reactive oxygen species [174].

The most controversial data relates to the correlation of antioxidant content with age. Many studies have been performed, but the work of R. G. Cutler or R. Sohol are the most impressive [42, 43, 180, 181]. Cutler developed the constants called the mean life-span potential (MLSP) and the specific metabolic rate (SMR). Generally the correlation between these two constants forms a hyperbola with man in an extreme position. The SMR is proportional to the rate of oxygen utilization in tissues and with the formation of reactive oxygen species. Therefore, the ratio of antioxidant content and SMR should indicate the degree of protection of an organ against reactive oxygen species produced during metabolism. If reactive oxygen species are involved in the aging process there should be a positive correlation between MLSP and various antioxidants. Indeed, for the common house fly a linear correlation is found between MLSP and the ratio of vitamins C or E to SMR and for the ratio of glutathione transferase and daily caloric intake. Professor T. V. Slater has postulated that organisms evolved along the MLSP line in order to keep the formation of reactive oxygen species within reasonable limits. This was accomplished by the inclusion of various types of enzymatic and nonenzymatic antioxidants among the basic components of the cell [177].

This point of view points to several tendencies that occur during the aging process:

- The gradual increase in reactive oxygen species formation.
- The gradual decrease of cellular antioxidants and repair enzymes [77, 181].
- Long-term exposure to prooxidant agents such as radiation or chemical pollutants [127].

The first two of these are partly the result or effect of the third. Not all organs of an animal are equally resistant to oxidative stress, so there are critical organs to this process where a disease may arise. The overcoming of the antioxidant defenses by the first and third influence the individual rate of aging. This rate can be measured by several methods.

One method is the measurement of the amount of daily release of DNA end products in the urine, such thymine glycols or 5-hydroxymethyl uracil (table 8.3). These end products are the result of hydroxyl radical attack on DNA [5, 41, 205]. Ames and his coworkers have shown that approximately one million cell mutations occur every day due to exposure to reactive oxygen species or mutagens [5, 142]. Most of these errors are repaired by the enzymes provided for this purpose, and the rest are destroyed and the end products released to the urine [5, 77, 115]. These products can be quantitated using high pressure liquid chromatography (HPLC). Studies by American and Japanese scientists have confirmed that these end products arise from destroyed DNA and not from the intestinal flora. The amount of daily excretion of these end products (table 8.3) indicates a mean of 103 molecules of thymine oxidized per day for each molecule of DNA affected [5].

Oxidation rate can also be determined by the measurement of the end products found in exhaled air, such as ethane or pentane. These products result from the breakdown of peroxides to aldehydes and then short chain hydrocarbons, which make their way to the lungs and are exhaled [142].

Electron spin resonance (ESR) can be used to measure free radicals in the organs of animals. However, this method is only available in the research setting.

The least expensive method of measuring the extent of damage due to oxidative stress is the measurement of lipofuscin pigments stored in various cells. This technique has been used to measure lipofuscin using micro-spectrofluorometry in myocardial and glial cells as a measure of senescence and oxidative stress [180, 181]. The involvement of reactive oxygen species in the formation of lipofuscin in cells is clearly demonstrated and relates to the metabolic rate or antioxidant content in the diet [4, 181]. This supports Harman's hypothesis of aging. Even critics of this hypothesis admit that reactive oxygen species are involved in degerative diseases related to aging, such as cardiovascular disease.

Table 8.3 Daily human urinary excretion of DNA breakdown products [5].

Compound	nmoles/kg	molecules/cell
Thymin glycol	0.39	270
Thymidine glycol	0.10	70
5-hydroxymethyluracil	0.90	620
5-hydroxymethyl-2'-deoxyuridine	trace	trace

Based on this hypothesis and related observations, it is suggested the effects of aging can be limited or slowed.

Experimental epidemiological studies show an inverse relation between caloric intake or body weight and life span. Moderate caloric restriction reduces metabolic rate and, consequently, oxygen consumption (mitochondrial respiration).

Dietary lipids strongly influence the polyunsaturated fatty acid content of triglycerides in the adipose tissue. Some polyunsaturated fatty acids cannot be synthesized by the human body so they must be supplied in the diet (essential fatty acids). The incidence of cardiovascular disease is influenced by the quantity and saturation of dietary lipids [123, 173, 202]. Oxidative stress is one factor implicated in cardiovascular disease.

Antioxidant intake, especially the use of supplementation, is a hotly debated issue regarding its effect on disease resistance. Because there are so many enzymatic and nonenzymatic antioxidant systems in the body, the effect of supplementation of any one or several antioxidants is difficult to clearly detect. In addition, it is pointed out that the excess antioxidant is simply excreted in most cases and not stored for later use. This is particularly true for vitamins C and E and sulphydryl group donors [14, 101, 199, 206].

The life-span of experimental animals, from insects to mammals, may be influenced by natural or synthetic antioxidants, such as DMSO, BHT, nordihydro-guaiaretic acid, or thiazolidin-4-carboxilic acid. However, complex problems arise with the repeated administration of these compounds, such as the decrease in the antioxidant content of the blood. As Cutler [41, 42] and Sohal [180, 181] have shown, some antioxidants are metabolized and some of this metabolism can produce free radicals. This will logically result in a temporary decrease in total antioxidant content. In addition, antioxidant rich foods contain other factors that stimulate antioxidant metabolism. A third factor is that when the antioxidants in the diet is increased the tissue content returns to a normal physiological level following an adaptive period due to homeostatic mechanisms [8, 107, 150].

According to Sohal [180, 181], antioxidant supplementation increases MLSP in many species, including man. For example, the administration of a nontoxic specific inhibitor of superoxide dismutase (diethyldithio carbamate) decreases the enzyme activity. However, there is a compensatory increase in the glutathione level, and the MLSP is increased. Similar results were obtained following the administration of a specific inhibitor of catalase (3-aminotriazol) or by moderate oxidative stress induced by low concentrations of diamide or paraquat [121, 168].

A significant problem in aging is the effect on the brain. For both humans and animals, the aging process starts early in the brain. The brain is rich in polyunsaturated fatty acids such as 22 ω3 (cervonic acid) and 20 ω6 (arachidonic acid) making it an ideal area for peroxidation. In addition, the brain is relatively poor in antioxidant systems as its main line of defense is the blood-brain barrier and modest amounts of superoxide dismutase, ascorbate, and melatonin. Also, the brain contains iron deposits, especially in the basal ganglia. Therefore, when oxidative stress occurs in the brain (trauma or hemorrhage) a favorable environment for oxygen activation and peroxidation may arise [34, 75, 174]. The most sensitive areas of the brain have been determined experimentally to be the cortex > cerebellum > hippocampus. The least sensitive are the striatum, hypothalamus, and brain stem [32, 79, 199].

The superoxide dismutase concentration within the brain varies. The most is found in the hypothalamus followed by the cortex and the rest of the brain. With aging, the amount of superoxide dismutase and glutathione reductase decrease while activity of the hexose monophosphate shunt increases to compensate. The use of a calcium blocker (nicardipin) demonstrated that the effects of free radicals can be reduced. Harman also found a correlation between the formation of lipofuscin pigments in different portions of the brain and the localization of cerebral lesions or stroke or cerebral atherosclerosis [43, 77, 203].

In conclusion, free radicals are clearly involved in aging. Stressful conditions (chronic infection, trauma, emotional stress, exposure to pollutants) may accelerate the aging process. We still do not have a clear understanding of how much supplementation with antioxidants slows this process.

8.5 HYPERBARIA

Chronic or acute intoxication with oxygen has been demonstrated [29, 102]. The toxic role of oxygen is of increasing interest and concern as pure oxygen is used in aviation, space flight, and medical therapy. The biological mechanism of chronic oxygen intoxication involves the formation of large amounts of reactive oxygen species and peroxides. Most studies show that the erythrocytes, lungs, and brain are the main targets of reactive oxygen species, but the consequences are very different.

Table 8.4 summarizes the effects of hyperbaric oxygen on some parts of the body. These modifications suggest the critical organs are the brain (by alteration of the blood-brain barrier and increased reactive oxygen species formation) and lungs (activation of leukocytes and release of reactive oxygen species) [29, 102].

While superoxide dismutase is activated, at least in leukocytes, the administration of purified superoxide dismutase in liposomes or aerosols did not provide conclusive results because of the influence of the blood-brain barrier. However, the administration of free sulphydryl group donors (glutathione, N-acetylcysteine) or succinate (12 nmol/kg) produced promising results. The role of succinate is interesting because glutamic acid is converted to GABA which is converted to succinate semialdehyde by GABA transaminase which is in turn converted either to succinate or GOPA, depending on relative concentrations. Glutathione regenerates glutamate by a special pathway called the glutathione-GABA-succinate pathway, which seems to be used under stressful and hyperoxic conditions. In such conditions, the maintenance of physiological levels of mitochondrial respiration and tricarboxylic acid cycle activity are essential for survival [32, 118].

Table 8.4 Biochemical and clinical changes under hyperbaric conditions.

Organ	Species	Effect
Liver	Mouse	Decrease in oxygen consumption and of succinate dehydrogenase activity
Brain	Rat	Decrease of mitochondrial respiration and of a-ketoglutarate dehydrogenase
Erythrocytes	Man	Decrease of resistance to oxidative agents, increased met-hemoglobin, hemolysis
	Rat	
Leukocytes	Man	Increase of oxygen consumption, reactive oxygen species formation, and superoxide dismutase activity and decrease in catalase and glutathione peroxidase activity
Macrophages	Rat	

8.6 CHEMICAL POLLUTION

Under the conditions found in industrialized countries, the main human exposure to exogenous sources of free radicals is chemical pollution [127]. The amount of pollution in our environment and the nature of atmospheric and water circulation means that everyone is exposed to the toxic effects of pollution whether they live close to or far from an industrial source. In some parts of the world the amount of pollutants in the air, water, soil, and food threaten the health and survival of humans, plants, and animals. Most pollutants that we come in contact with produce free radicals either directly or during metabolism. Therefore, the main defense against these effects are the nonenzymatic and enzymatic antioxidant systems.

The main risk of exposure to pollution is the cumulative effects of free radicals over time. However, there also severe consequences to higher, acute exposures to free radical producing pollutants. The end result is the formation of oxidative stress in critical organs with the production of reactive oxygen species and serious disease. The long incubation period for the development of cancer and other diseases resulting from exposure to pollutants is due to the presence of the antioxidant systems that take time to wear down. Also, much of the low-level or chronic damage caused by exposure to pollutants is repaired by the body before it can lead to disease.

The critical target in the body are the nucleic acids that make up the genetic information of the cell. The formation of DNA-adducts with pollutants or their metabolites produces structural modifications, some of which are mutagenic [88]. The toxicity of pollutants is expressed as the affinity constant, Ks. The most toxic and mutagenic compound known is dioxin with a Ks or 1 nM [5, 58, 137, 146].

The binding of a pollutant to DNA modifies the activity of the genetic material, allowing activation and transcription of structural genes for mono-oxygenases, which are cytochrome P_{450} dependent enzymes. These new, modified cytochromes ($P1_{450}$, P_{448}) transform chemical pollutants into activated metabolites with increased carcinogenic potential. Sensitive tissues, such as smoker's lungs, which come into contact with tar and aromatics from smoke, produce appreciable amounts of cytochrome P_{450} variants [100, 176, 204, 207]. Lymphocytes and monocytes isolated from blood and exposed *in vitro* to an inducer (3-methylcholatrene) results in the induction of a specific hydroxyl compound that increases the risk of lung cancer. This is the basis of a test that may be useful to establish individual sensitivity to pollutants [137, 162, 191].

There is now evidence that individuals can adapt to oxidative stress. This process is a very important one for the involvement of free radicals in biology and medicine. A wide variety of compounds, including oxygen, cause oxidative stress if they are present in sufficient concentration. Wiese recently demonstrated that mammalian cells (fibroblasts, liver) can adapt to a fluctuating exposure to hydrogenperoxide in a wide range of concentrations without showing morphological changes [201]. The mechanism of this adaptation is varied and includes transient over expression of genes for antioxidant enzymes, increased synthesis of shock proteins, temporary growth arrest, and lengthening of the cell cycle. Only high concentrations of hydrogenperoxide (30 μmol/10^7 cell or about 30 mM) resulted in cell death.

Adaptive processes affect all levels from cells to the entire body. Some examples are natural processes (pregnancy, birth, exercise) while others are imposed on the organism from outside (pollution). In all these cases, the response is the increased synthesis of antioxidant systems. However, such an adaptation cannot continue indefinitely. Long-term exposure to oxidative stress disrupts the dynamic equilibrium of the organism. This may lead to decreased resistance to infection. Trauma may trigger a pathological condition that is usually localized to a sensitive organ [127].

Chapter 9

FREE RADICALS AND PATHOLOGY

To demonstrate the involvement of free radicals in aging and various pathological conditions is not a simple or easy job. We have indicated that oxygen activation is an ongoing event and that the body has elaborate defenses against the harmful effects of reactive oxygen species and peroxides. We have indicated that the formation of reactive oxygen species or peroxides have some beneficial effects. We have indicated that mild oxidative stress does not lead to disease if the antioxidant and repair systems of the organism are functioning and are not overwhelmed. We will now take a closer look at specific pathological conditions in which free radicals are involved.

9.1 INFLAMMATION AND RHEUMATIC CONDITIONS

Acute inflammation is possibly the most widespread pathological condition found, ranging from the common whitlow or granuloma to the irreversible morphological alteration of tissues or organs. Acute inflammation is very complex, consisting in varied sequences of biochemical, biological, and hematological process that are not entirely understood [38, 125, 200]. This complexity is illustrated in figure 9.1. Acute inflammation following microbial infection, trauma, or irritation from a foreign compound occurs locally and is usually limited to a certain tissue. Regardless of the cause of the acute inflammation, the first steps are the same. Later steps vary depending on the nature of the insult [70, 89].

Also shown in figure 9.1 is the point that as inflammation is developing there is a tendency for more tissues and biological systems to become involved. This is part of the cause of the complexity of inflammation. The following points should be made about inflammation.

A. Acute inflammation induces a strong and varied activation of the body's defenses.

- Mast cells release mediators such as histamine
- Plasma kinogenases activate kinins
- Biosynthesis of prostaglandins, leukotrienes, and thromboxanes is stimulated

- Infiltration by phagocytic leukocytes and activation of leukocytes with the release of reactive oxygen species, proteases, and hydrolases
- Reactive oxygen species have a direct or indirect (activation of damaging enzymes) effect on cell membranes causing increased permeability and peroxidation
- Stimulation of phagocytic leukocytes will induce the activation of phospholipase, releasing arachidonic acid from membranes, stimulating prostaglandin biosynthesis

B. Experimental models of inflammation, especially acute inflammation, indicates two distinct phases are present.

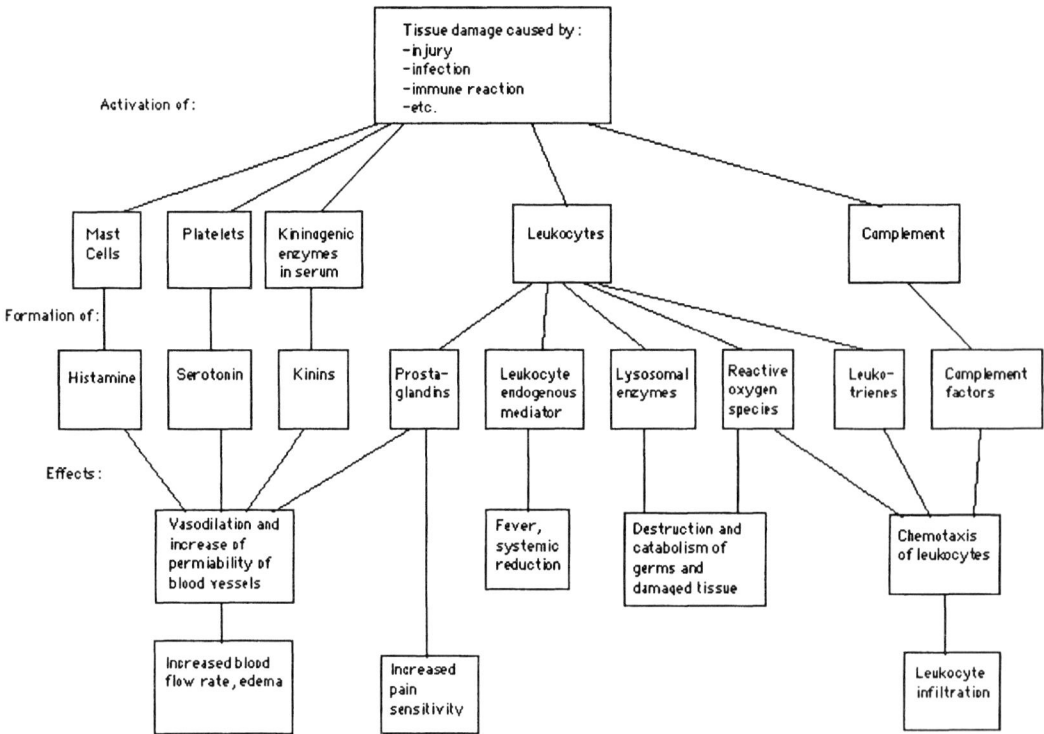

Figure 9.1. Local inflammatory reactions with classical symptoms of inflammation (dolor, rubor, calor, tumor = pain, redness, warmth, swelling) are explained by vascular effects and leukocyte infiltration.

The first phase leads to the formation of edema is due to the combined effects of histamine, serotonin, and kinins, and produces increased permeability and dilation of the blood vessels

The second phase is regulated by arachidonic derivatives (prostaglandins and leukotrienes), especially leukotriene B_4, which is the strongest chemotactic factor for phagocytic leukocytes

The sequence of reactions occurring in acute inflammation is a marvel of complexity and logic. Indeed, phagocytic leukocytes (neutrophils, macrophages) release both chemotactic and antichemotactic factors, vasodilators (PGE_2) and vasoconstrictors

($PGF_{2\alpha}$) to regulate or limit their actions. This regulation consists of several branches of reactions and is important because of the great variety of compounds that are found at the site of acute inflammation, some of which have contradicting effects. Arachidonic acid and its derivatives play a special role in this activity. Most antiinflammatory drugs influence the metabolism of arachidonic acid by preventing the formation of the endoperoxide (PGH_2), which is a reaction catalyzed by cyclooxygenase.

C. The increased formation of reactive oxygen species starts with the first step of acute inflammation and rises with the activation and mobilization of leukocytes, as reactive oxygen species are used to kill invading pathogens. The increased formation of reactive oxygen species also induces the following.

- The release and peroxidation of polyunsaturated fatty acids from membranes, followed by prostaglandin biosynthesis
- Structural modifications of hyaluronic acid and collagen from the matrix of neighboring tissues
- Structural degradation of DNA and polynucleotides
- The formation of edema and of secreted mediators
- Increased bactericidal cytolytic activity, directly or mediated by activated lymphocytes
- Activation of peroxidase, hydrogenperoxide, and halogens for bactericidal activity, the cause of chemoluminesence during leukocytic phagocytosis. This is the most convenient test for measuring an inflammatory condition [125, 156].

The formation of peroxides has a feedback effect by inactivating chemotactic factors released at the beginning of acute inflammation. This tends to regulate the extent of this process.

D. Autoregulation and the varied activation of defense systems may be explained by the exchange of chemical signals when cellular and humoral agents are released, as suggested in figure 9.2. Note that at the beginning of acute inflammation humoral factors prevail as leukocytes are mobilized and activated.

E. During acute inflammation, the adhesion of phagocytic leukocytes to the walls of the vascular endothelium occurs. Activated leukocytes adhere strongly to the vascular bed. Factors involved in this are listed in table 9.1. The adhesion of leukocytes to endothelial cells lining the walls of blood vessels is a major source of plaque formation and a risk factor for cardiovascular disease [11, 54, 89].

F. When acute inflammation spreads, a systemic reaction takes place due to the inflammatory reactions affecting the entire body.

- Increased circulation of leukocytes from stimulation of bone marrow

- Increase of aminoacids in the blood, which are required for intense protein biosynthesis
- Decrease of plasma concentrations of albumin and transferrin
- Decrease of plasma concentrations of zinc and iron due to their absorption by the liver
- Increased plasma level of acute phase proteins such as haptoglobin, C-reactive protein, fibrinogen, and ceruloplasmin. The last two proteins are also antioxidants.
- Increased plasma level of copper in the free state or bound to ceruloplasmin. Remember that copper ions together with hydrogenperoxide (released from activated leukocytes) is a powerful reactive oxygen species generating system. The result is the inactivation of proteins such as $\alpha 1$-proteinase inhibitor. As a result, elastase is free to degrade connective tissue, which occurs in rheumatoid arthritis [40, 182].

Table 9.1 Factors that influence neutrophil adhesion

Stimulators	Inhibitors
Plasma from patients with AI	Plasma from patients treated with antiinflammatory drugs
cGMP	cAMP
Propranolol	Adrenalin (epinephrine)
Chemotactic factors*	Colcuicyn
Calcium*	Prostaglandins *
Reactive oxygen species*	Antiinflammatory drugs*

*Modifications related to reactive oxygen species actions

G. Acute inflammation should be regarded as a complex defense reaction of the whole organism. In the majority of cases, complete recovery occurs. However, there are also cases in which an inflammatory site remains localized in the tissue where it first developed or in another area of the body. This is a chronic inflammatory reaction.

The anomalous increase of inflammatory reactions, especially of phagocytic leukocytes, is the main source of several diseases such as rheumatism and autoimmune diseases. One of the marked features of rheumatoid conditions is the increase of synovial fluid. In rheumatic conditions, the composition of the synovial fluid is modified: viscosity is decreased (due to altered hyaluronic acid), iron and peroxides increase, leukocytes are attracted, antioxidant enzymes such as superoxide dismutase and catalase decrease. The measurement of iron in synovial fluid seems to be a reliable criterion for the patient's prognosis because it decreases in efficient treatment [10, 166, 189].

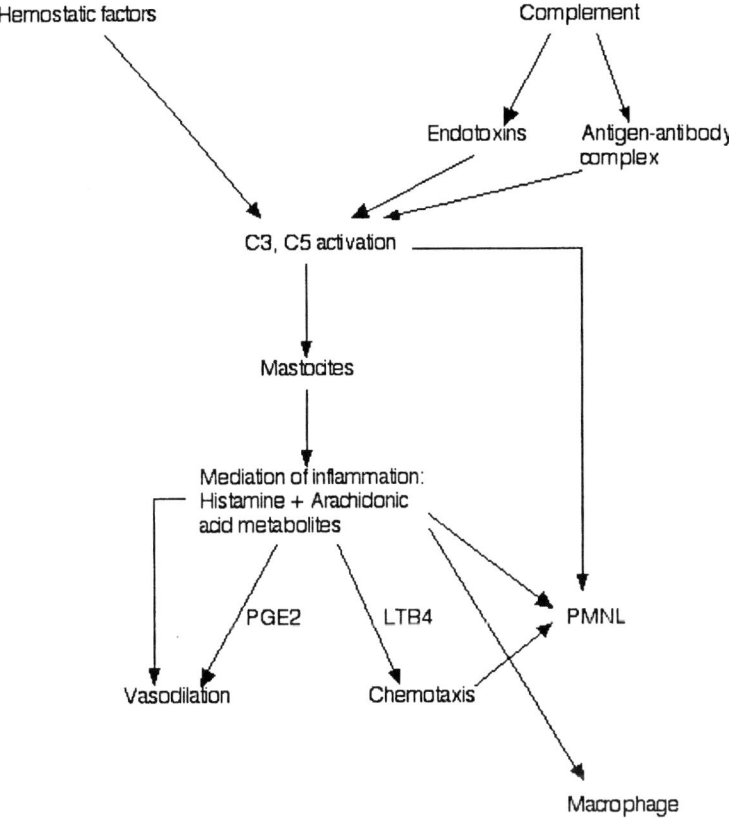

Figure 9.2. Schematic representation of some humoral and cellular immune factors activated during inflammation and their effects.

H. The most impressive involvement of reactive oxygen species in the inflammatory process is the action of antiinflammatory drugs. These drugs are not structurally related and act by several mechanisms, although almost all anti-inflammatory drugs function by inhibiting or altering the biosynthesis of eicosanoids (prostaglandins and related compounds).

As discussed in chapter 4, the involvement of reactive oxygen species in eicosanoid biosynthesis starts with the release of arachidonic acid. As shown in table 9.2, eicosanoids affect virtually every part of the body, and alterations in their synthesis can therefore have wide ranging and unexpected consequences. Leukotriene B_4 (LTB_4) is a proinflammatory agent as it is chemotactic and favors the aggregation of platelets. Prostaglandin E_2 (PGE_2) is antiinflammatory because it decreases phagocytosis and the release of calcium ions.

Many experimental studies, both *in vivo* and *in vitro*, have shown that most antiinflammatory drugs block one or both biosynthetic branches of eicosanoids, as do the natural antiinflammatory compounds.

Table 9.2 The biological effects of eicosanoids (derivatives of arachidonic acid)

Biosynthesis by cyclooxygenase

Tissue or condition	Effect
Blood vessels	Vasodilation: PGE_2, PGI_2
	Vasoconstriction: TXA_2
Lung	Bronchioconstriction: $PGF_{2\alpha}$, PGD_2, TXA_2
	Bronchiodilation: PGE_2, PGI_2
Platelets	Aggregation: TXA_2
	Antiaggregation: PGI_2
Stomach	Increased acid secretion with decrease of PGE_2, PGE_1
Intestine	Contraction of longitudinal muscles: PGD_2, TXA_2, $PGF_{2\alpha}$
	Contraction of circular muscles: $PGF_{2\alpha}$
	Increased water and ion secretion: PGE_2
Adrenals	Secretion of corticosteroids: PGE_2
	Secretion of aldosterone: PGE_2
Growth and lactation	Increase secretion of growth hormone and prolactin: PGE_2, $PGF_{2\alpha}$
Pain	Increased by PGE_2, PGD_2, PGI_2
Fever	Reduced by PGE_2, PGI_2

Biosynthesis by lipoxygenase

Tissue or condition	Effect
Smooth muscle	Vasoconstriction: LTC_4, LTD_4
Blood vessels	Increased permiability: LTC_2, LTD_4, LTB_4
Phagocytic cells	Chemotaxis: LTB_4, 5-HETE, 9-HETE, 11-HETE, 12-HETE
Endothelium	Increased leukocyte adhesion: LTB_4
Stimulates secretion	Lysogen by neutrophils, histamine by basophils, insulin by pancreatic beta cells: LTB_5, 5-HETE, 12-HETE

- Superoxide dismutase blocks endoperoxide (PGH_2) formation
- Glucocorticoids (hydrocortisone) and mepacrine block phospholipase A2, inhibiting the release of arachidonic acid from membranes
- Nonsteroidal antiinflammatory drugs (NSAIDs), such as aspirin, indomethacin, and phenylbutazone, inhibit cyclooxygenase
- Synthetic antioxidants, such as propyl gallate and nordihydroguairetic acid, selectively inhibit lipoxygenase catalyzed reactions
- Diclofenac and other antiinflammatory drugs inhibit both branches of eicosanoid biosynthesis.

Most antiinflammatory drugs act on several sites, including their action of eicosanoid biosynthesis. For example, glucocorticoids may influence over 15 processes or enzyme reactions. All antiinflammatory drugs have some effects in common: inhibit eicosanoid biosynthesis, have antioxidant activity, act on membrane permeability by blocking the release of hydrolytic enzymes (proteases).

I. A very interesting field with large prospects for therapy is the role of copper in inflammation and degenerative diseases. In these pathological conditions the copper level in the plasma increases from 1 µg/ml to 2.5 to 3.0 µg/ml. Copper binding proteins, such as ceruloplasmin and albumin bound copper also increase [40, 182].

Interleukin IR-1 triggers the release of copper from liver deposits, which is released as copper-metallothionein into the blood. In several pathological conditions, such as rheumatism and cancer, the organism requires copper to intensify metabolic activity. When the condition regresses, the level of plasma copper decreased to its normal limits. If copper or ceruloplasmin in the blood remains elevated, the prognosis is poor.

According to Sorensen, the variation of copper in the blood is more significant than other nonspecific modifications, such as hemoglobin and erythrocyte rate of sedimentation [182]. It was found that the increase in cupremia is due mostly to the loosely bound copper associated with albumin and amino acids, which can oxidize free sulphydryl groups in plasma, hemoglobin or lysosomal membranes within leukocytes. These modifications are in agreement with the magnitude of clinical symptoms in rheumatism. Therefore, the use of antiinflammatory drugs such as penicillamine or gold salts in rheumatism may be explained by the chelation of copper. The copper complex of antiinflammatory drugs (indomethacin) is more effective than the drug alone. The plasma of patients treated with antiinflammatory drugs have a greater total antioxidant content due to the increased level of ceruloplasmin. Organic complexes of copper that exhibit superoxide dismutase-like activity decrease cupremia in polyarthritis.

J. The association of chronic inflammation with cancer has been known for two centuries. Ulcerative colitis, gastric ulcer, and cholecystitis can lead to a malignant condition. Some researchers believe the mobilization and activation of phagocytic leukocytes at the site of a chronic inflammatory process will increase the release of reactive oxygen species [200]. This can interfere with the surrounding cells.

- Activate precancerous metabolites
- Increase the peroxidation and degradation of peroxides to mutagenic aldehydes
- Directly affect DNA and protein biosynthesis

All these processes favor structural modifications of DNA. This is, of course, a long process, which is favorable influenced by the presence of antioxidants [4, 88, 104, 136, 146].

9.2 CARDIOVASCULAR DISEASE

In industrialized countries cardiovascular disease is the major cause of death, and occlusion of the coronary arteries by atheromatous plaques accounts for most of this mortality. This fact is widely publicized and is sustained by experimental and epidemiological data. Demographic differences in the frequency of cardiovascular disease have been interpreted as due to life-style or dietary differences, especially those involving cholesterol. While most public health efforts to prevent heart disease are aimed at reducing cholesterol, the cause and effect relationship is not without problems [11, 151, 202].

The implication of free radicals or reactive oxygen species in the etiology or course of cardiovascular disease is sustained by a large array of analytical data and *in vitro* studies, but few clinical trials are available. The presence of an increased content of lipid peroxides in the plasma of cardiovascular disease patients has been reported since the 1970s [140, 203]. These observations have been confirmed by the demonstration of increased amounts of peroxides and lipofuscin pigments present in atheromatous plaques within blood vessels. Professor Yagi, from Nagoya University, made scanning electron micrographs that demonstrate the harmful effects obtained following administration of native or peroxidized linoleic acid on the endothelial surface the arteries. Ulceration and morphologic lesions were produced only following administration of peroxidized linoleic acid. Atheromatous plaques were also greatly increased in their content of native and oxidized cholesterol (10 fold increase over noninvolved areas) and peroxidized lipids [203].

The risk factors for cardiovascular disease include smoking (89%), hypertension (78%), diabetes mellitus (25%), and hyperlipemia (20%). To these independent biological risk factors are added factors like increased white cell count and increased fibrinogen level [11]. All the above factors favor the formation of oxidative stress triggered by chronic infection, inflammation, or a blood clot adhering to a vessel wall. These events trigger the accumulation of phagocytic leukocytes and platelets, which adhere to the endothelial layer of the blood vessels. Activated leukocytes, macrophages, and endothelial cells are a great source of reactive oxygen species, which will peroxidize cell membranes and lipoproteins (especially LDL). There is abundant literature that demonstrates that LDL easily becomes peroxidized. Antibodies to peroxidized LDL have been identified in the blood of patients with cardiovascular disease. LDL is easily peroxidized due to its content of polyunsaturated fatty acids. Peroxidation gives LDL an electronegative charge that is recognized by the scavenger receptors on the surface of macrophages. These scavenger receptors are specifically blocked by polyinosine and polyguanilic acid. The engulfing of oxidized LDL by macrophages changes their morphological structure, leading to the formation of foam cells.

Cytokines released by activated macrophages stimulate collagen secretion and smooth muscle proliferation [11, 151, 179, 196, 209]. In many tissues, this process leads to healing. However, in the walls of the coronary arteries, the resulting scar may have catastrophic consequences as it can lead to myocardial infarction or ischemia [58, 66].

A diet rich in antioxidants should reduce LDL oxidation, as has been demonstrated experimentally with animals [94, 107, 119, 202, 205]. Only lipid soluble antioxidants are able to effectively inhibit LDL oxidation. In humans, LDL cholesterol contains approximately 4 nmol/mg total antioxidant (2.8 nmol/mg vitamin E, 0.4 nmol/mg carotenoids [including β-carotene and lycopene], and 0.6 nmol/mg retinoid esters. But, under physiological conditions, ascorbic acid may also inhibit LDL oxidation. The amount of polyunsaturated fatty acid present in the LDL strongly influences the yield of oxidation. It was found that the ratio of antioxidant to polyunsaturated fatty acid in LDL is 1 to 170. Therefore, LDL particles behave like easily oxidized unsaturated fatty acids.

After LDL peroxidation, the same decomposition products as occur with polyunsaturated fatty acid peroxidation appear (aldehydes, malondialdehyde, 4-hydroxynonenal). These aldehydes react with the amino acids of some proteins, such as apolipoprotein A, which allows the LDL to bind to the scavenger receptor of macrophages. Monoclonal antibodies towards malondialdehyde or other aldehydes recognize oxidized LDL. Since the scavenger receptor of macrophages is unregulated, unlimited uptake of oxidized LDL results, causing the conversion of the macrophage into a foam cell.

Vascular endothelium is the cell monolayer that covers the interior of the blood vessels. The endothelial cells form a continuous covering of the 50 km of blood vessels. Thrombosis takes place at the level of the endothelial cells following scaring or inflammation. The sequence of events that seems to take place are as follows.

- Mobilization of platelets and leukocytes to the endothelium
- Aggregation of these cells to the endothelium
- Aggregated platelets release factors that stimulate coagulation
- Endothelial cells release hydrogenperoxide when irritated
- Fibrinogen is broken down to fibrin by activated thrombin. The clot includes fibrin and cells [11, 100, 132, 165, 179, 187, 209].

Endothelial cells resist platelet aggregation by releasing inhibitors such as EDRF (endothelial-derived relaxing factor), now know to be nitric oxide (NO). This vasodilating compound is synthesized from arginine by the enzyme nitric oxide synthase, a heme-containing flavoprotein.

Arginine — synthase   citrulline + NO

Peroxidized LDL inhibits the synthesis of NO and alters the balance between prooxidants and antioxidants.

The action of EDRF (or NO) is still only partly known, but it is involved in relaxing the smooth muscles of the blood vessels and the inhibition of LDL oxidation, the release of chemotactic and cytotoxic factors. Thus, NO acts both as a chemical signal and as a weapon that enhances local defenses. Most antioxidants act as prooxidants under the right conditions. This is also true for NO when reactive oxygen species are already present in the area. Acute inflammation releases large amounts of reactive oxygen species, which involve NO in a prooxidative reaction with hemoglobin.

Myoglobin + H_2O_2 Met-myoglobin

Met-myoglobin is a free radical that accelerates LDL oxidation.

Recent studies have examined the bactericidal properties of nitric oxide [145]. In humans, increased salivary nitrite production resulting from nitrite intake enhances oral NO production. The conversion of nitrite to NO may be due to acid producing bacteria.

Many microbes, such as Clostridia or Mycobacterium, are susceptible to the killing action of NO.

As seen in table 9.3, NO can have opposing effects in the same tissue. Nitric oxide contains an extra electron, thus making it highly reactive with a variety of different molecular targets, especially proteins, thiols, and heme-containing compounds. Nitric oxide has a half-life of seconds. The biochemical and medical applications of NO are still being studied.

Table 9.3 Roles of nitric oxide, after Redomski [145].

Tissue	Role	Toxicity
Blood vessels	Ischemic protection, antiatherosclerotic, antiadhesion	Septic shock, inflammation favors atherosclerosis
Heart	Negative ionotropic ischemia	Myocardial stunning, septic shock
Lung	Ventilation-perfusion, mucus secretion, immune defence	Immune complex-induced alveolitis
Kidney	Tubuloglomerular feedback	Glomerulonephritis, acute kidney failure
Central nervous system	Synaptogenesis, enhanced memeory formation, cerebral blood flow, neuroendocrine secretion, visual transduction, olfaction	Neurotoxic, proconvulsive, migraine, hyperalgesia
Pancreas	Endocrine / exocrine secretion	Beta-cell destruction
Gut	Blood flow, peristalsis, exocrine secretion, mucosal protection, antimicrobial	Mutagenesis, mucosal damage
Immune system	Antimicrobial, possibly antitumor	Antialograft, inflammation septic shock, tissue damage

Hypertension occurs when there is increased resistance to blood flow, often due to vasoconstriction. We will avoid a detailed discussion of hypertension, and will only mention that in the elderly a progressive arteriocapillary fibrosis appears that is associated with an increase in lipid peroxides, copper, and iron in the blood. At the same time, free sulphydryl groups and vitamin E decreases. In pregnancy there sometimes occurs a toxic hypertensive condition associated with hyperlipidemia and a decrease in vitamin C in the blood. In women taking oral contraceptives, a temporary hypertension may appear that is associated with hyperlipidemia [39, 54, 150, 202].

In these cases of hypertension, an association with hyperlipidemia favors increased lipid peroxidation. As was mentioned in chapter 4.4, an increased level of peroxides triggers a shift in eicosanoid biosynthesis towards thromboxane (TXA_2) and endoperoxides (PGG_2), which damage the vascular walls. Such a mechanism was demonstrated experimentally on animals and has been observed in heavy smokers and cases of endotoxin intoxication. Such modifications favor an increased aggregation of platelets [165]. As prostacyclin inhibits platelet aggregation and vasoconstriction, the increase of plasma lipid peroxidation should inhibit prostacyclin synthase, especially in the endothelium. Unpublished data from Dr. Olinescu's laboratory that relates to this observation and involving 700 patients who underwent cardiac catheterization is presented in table 9.4.

Table 9.4 Lipid peroxides (TBARS*) in the plasma of patients with cardiovascular disease, sorted by age group. After Olinescu [to be published].

Condition	n	Age group (years)		
		25 - 50	51 - 69	70+
Control	60	2.49±0.16	2.66±0.27	2.73±0.32
Moderate stenosis	286	3.05±0.26[a]	2.82±0.23	3.24±0.11[a]
50% stenosis	80	3.08±0.18	3.24±0.88[a]	3.82±0.24[a]
100% stenosis	434	4.43±0.32[a]	4.27±0.27[a]	4.12±0.34[a]
Hypertension	120	2.53±0.32[a]	3.45±0.26[a]	4.09±0.32[a]
MI survivors	40	3.16±.021[a]	3.28±0.16[a]	4.32±0.12[a]
Smokers	150	2.78±0.19	3.46±0.18	3.87±0.14[a]

*Thiobarbituric acid reactive substances
[a]Significantly different from control

The data in table 9.4 shows that plasma lipid peroxidation increases with age. However, in almost all other groups, lipid peroxide is significantly increased for the same age, especially for the oldest groups with significant stenosis. This study emphasizes the role of aging in the etiology of cardiovascular disease, as well as the role of oxidative stress. The presence of oxidative stress is shown by the increased values for lipid peroxidation, fibrinogen, ceruloplasmin, and cholesterol (in the elderly) and especially the leukocyte activation while protein thiols and total antioxidants are decreased.

This study shows that coronary vascular disease is more or less connected with aging, and an inflammatory condition (infection, smoking, stress) was also found in most cases. A chronic inflammatory condition is a powerful source of reactive oxygen species for the activation of leukocytes.

It was noticed many years ago that coronary vascular disease involves significant modification of lipids, cholesterol, and lipoproteins. LDL cholesterol has a particularly strong position to play in this condition.

The characteristics of lipoproteins are given in table 9.5. It is quickly apparent that LDL contains high amounts of cholesterol, phospholipid, and polyunsaturated fatty acid. Therefore, this lipoprotein fraction is a particularly suitable substrate for peroxidation. The actual risk factor is the LDL cholesterol, but its quantitative measurement is expensive and time consuming. Therefore, total cholesterol is measures for screening purposes. Numerous studies have concluded that:

Table 9.5 The characteristics of plasma lipoproteins

Property	Fraction			
	Cholymicrons	VLDL	LDL	HDL
Density	1.006	0.98-1.00	1.00-1.063	1.10-1.20
Dimensions (Å)	2,000	500-1,000	100-300	70-100
Protein (%)	0.5	2-13	25-35	40-55
Triglyceride (%)	90	64-80	10-15	5-10
Cholesterol (%)	5	8-13	35-45	7-20
Phospholipid (%)	4	6-15	20-30	15-30
Cholesterol : lipid ratio	1.5	1.4	1.4	0.6

- In women, the greatest risk factors are high levels of triglycerides and low HDL.
- In men, the risk factors are, in decreasing order, hypercholesterolemia, smoking, hypertension, and diabetes.
- The level of HDL is increased by estrogens, a moderate alcohol intake, and physical exercise. The level of HDL is decreased by androgens and smoking.
- Various diets are recommend to decrease LDL cholesterol, but these diets tend to also reduce HDL.
- A diet high in omega-3 fatty acids (which are high in ocean fish) is believed to be responsible for the Eskimos very low incidence of heart disease. However, this diet has serious side effects, greatly increasing the time required for blood to clot. As a consequence, Eskimos eating a traditional diet are very susceptible to bleeding to death even from minor wounds.

Antioxidants and cardiovascular disease is another area that has drawn much research attention. The results are controversial. The study of Freddy showed an interesting relationship between the level of vitamin E in the plasma and mortality by cardiovascular disease in different countries. In Finland, Denmark, and England the content of vitamin E is about 24 µM. In mediterranean countries (Spain and Italy) the vitamin E content is nearly double. The higher the amount of vitamin E, the lower was the mortality from coronary vascular disease [63]. The multinational WHO-MONICA project, which lasted three years and involved scientists from seven European countries, found a clear relationship for the risk factors in human atherosclerosis (table 9.6). The data in table 9.6 emphasizes the role of vitamin E and vitamin C as antioxidants that can block the peroxidation of LDL.

Table 9.6 Pearson correlation coefficient (r^2) for atherosclerosis as determined by the WHO-MONICA Project

Parameter	r^2	Probability
Total cholesterol	0.04	0.53
Diastolic pression	0.08	0.80
Smoking	0.02	0.90
Blood vitamin A	0.22	0.13
Blood vitamin E	not listed	0.002
Plasma carotene	0.21	0.14
Blood vitamin C	0.41	0.03

A number of drugs can bring significant reductions in cholesterol levels and heart attacks. A University of Southern California study of men who had coronary by-pass surgery found cholesterol lowering drugs may reduce plaque inside arteries and raise the level of HDL. However, the long-term effects of cholesterol-lowering drugs have not been determined. Potential side effects of these drugs mean they should be used cautiously and under a physician's supervision [95].

The risk of cholesterol has been overstated. In the absence of smoking or high blood pressure, elevated total cholesterol alone does not appear to be a serious risk factor for coronary vascular disease. Low cholesterol, however, may have unanticipated side effects and adverse consequences such as stroke and cancer. In a study of 350,000 American men, ages 35 to 57, those with cholesterol levels below 160 mg/dl had a three times greater chance of suffering a hemorrhagic stroke, but less chance to develop coronary vascular disease [95, 151]. Cholesterol and fatty acids are also necessary for health. Lipids in lipoproteins and food contain vitamin E to protect against oxidation [71, 100, 170, 202].

Obesity is a risk factor for atherosclerosis, but this relationship is not a direct one. Several studies have shown that young men (in their twenties) who were in apparently good health had plaque-related lesions in their arteries, while many obese individuals have only age related amounts of plaque [11,141]. In America, weight loss is a national obsession. It should be as at least one-third of all Americans are over weight. More than $30 billion per year are spent by Americans to lose weight.

Obesity is the cause of many serious health problems including diabetes (adult onset or type 2) and coronary vascular disease. Most of the extremely obese have a hereditary condition. The genetic link to obesity was demonstrated in 1995 when American scientists demonstrated the existence of obese genes (ob) in mice. Mice homozygous for an defective ob gene are very obese, but when injected with protein coded for by that gene their weight dropped 40%. This establishes that these can be a genetic propensity for high body weight. Fatty infiltration of the liver and other tissues (which occurs with obesity) does not always result in hypercholesterolemia. In order to compensate for fatty infiltration, the number of peroxisomes and β-oxidation of fatty acids increase significantly. The level of lipid peroxides is only slightly increased compared to normal weight people. Some plasma antioxidants, such as ceruloplasmin or glutathione dependent enzymes are increased. These adaptive modifications provide protection for a limited time. They do not persist for the long-term because of the oxidative stress effects of aging [47].

9.3 CANCER

Statistical data from all industrialized countries indicates that cancer is the second leading cause of mortality. One view is that the increase in cancer in industrialized countries is due to elevated exposure to radiation and chemical pollutants [4, 37, 93, 113, 127, 164]. For the purpose of this discussion, we will assume that this observation is relevant. It is clear that both ionizing radiation and chemical pollutants cause cancer under experimental conditions. Both agents produce oxygen activation with the release of large amounts of reactive oxygen species. The exposure of humans to ionizing radiation or aromatic compounds has also been shown to result in a greater risk for developing cancer [197, 198].

The involvement of free radicals or reactive oxygen species in the etiology of cancer is a long debated problem. In some areas of oncology, such as chemical carcinogenesis, irradiation, or some cytostatic drugs (adriamycin, daunomycin) the involvement of reactive oxygen species has been demonstrated and is accepted as a side effect. The acceptance of free radical's implication in the etiology of cancer will only be accepted when the mechanisms of cellular differentiation and proliferation and control of cell division are understood [136, 177].

Among the hypothesis and suggested mechanisms Singh and Johal [176] have proposed a model of carcinogenesis consisting of several steps. Each step includes one of two processes. These are reversible up to a certain point as the organism exerts defensive mechanisms or through the effect of life-style or diet. That such a model represents real events is supported by the observation that a lengthy period of chronic exposure to a carcinogen is required before humans develop clinical disease. Even acute exposure may require an incubation period before disease develops. This is evident in the case of atomic bombing survivors in whom leukemia did not appear until 10 to 30 years after the event. A similar story is unfolding following the accident at Chernobil in the Ukraine in 1986.

It is also known that in experimental carcinogenesis the pre-cancerous lesions possess a degree of reversibility, depending on the degree of exposure to the inducer. This reversibility may be explained by the efficiency of specific or nonspecific defence systems that protect or repair the cell. Many of these are antioxidants [112, 115, 139, 159]. Hundreds of papers have been written that deal with the possible relationship between the frequency of cancer and dietary antioxidant content. If gastric dysplasia, cervical dysplasia, bronchial metaplasia, or urinary bladder papilloma as precancerous lesions, the assertion is supported as these are positively affected by a diet high in antioxidants [4, 22, 37, 110, 113, 137, 162].

It is still not entirely clear as to the initial step in the development of cancer. Most of the observed modifications, such as alterations in glycolysis, are a consequence of tumor growth. Therefore, the implications of reactive oxygen species in cancer are very hard to demonstrate. Due to metabolic alterations occurring in the host organism, the internal medium is increasingly prooxidative, emphasizing the alteration of genetic expression. This may explain the influence of promoters such as ionizing radiation and chemical pollutants. Polyaromatic hydrocarbons contain in their structure electron dense regions. Upon metabolism, these produce epoxides, free radicals, and other activated derivatives that bind DNA and induce structural alterations and mutations [51, 82, 104, 117, 200]. In addition, it seems that early in carcinogenesis, some endogenous compounds that may up-regulate cell growth are present in increased amounts. These include methyl glyoxal and other aldehydes [185] or polyamines, such as spermidine and putrescine. These compounds strongly affect DNA structure and function.

Cell proliferation is a very complex process and its fine regulation is the result of growth factors, cytokines, hormones, or second messengers that exchange signals between and within cells. These in turn activate specific genes, including proto-oncogenes, initiating a variety of key biochemical events. In all of this vast array of regulatory actions that take place just before or during proliferation, superoxide radical and hydrogenperoxide are acting by stimulating or inhibiting some reactions. These two reactive oxygen species stimulate the growth of cultured fibroblasts in a wide range of concentrations (10 nm to 1 µM). It is known that hydrogenperoxide can readily penetrate cells an possibly interact directly with transcription factors [2, 58, 109, 137, 148, 187]. Actually, both superoxide and hydrogenperoxide can stimulate growth in a considerable variety of cultured mammalian cell types when added exogenously. These two reactive oxygen species might function as mitogenic stimuli through biochemical processes connected to natural growth factors [136, 158]. Hydrogen peroxide in low concentrations activates transcription factor gene (egr 1) as well as c-jun and c-jos [163]. The addition of N-acetylcysteine counteracts this gene response.

Glutathione may play a role in cell division [7]. As soon as most cells (fibroblasts, lymphocytes) enter progressive growth, there is a decline in the amount of intracellular glutathione [6, 26, 122, 139].

Cell division involves multiple and impressive morphological changes involving the membrane and is likely to involve increased formation of superoxide and hydrogenperoxide, which may act as second messengers [27, 60, 189]. The implication of reactive oxygen species as second messengers for proliferation should be apparent when it is realized that chronic inflammation is a important source of these molecules. A relation between inflammation and carcinogenesis has long been suspected. Cells defend against increased reactive oxygen species production by activation superoxide dismutase and glutathione-related enzymes. Antioxidants from the diet may also play a role. This may help explain the long debated relationship between antioxidants and the frequency of cancer [3, 4, 18, 113, 200].

9.4. Respiratory Failure

The lungs are easily affected following exposure to oxidative gases such as oxygen, ozone, or nitrogen oxides at high concentrations, mineral dust (asbestos, silica), microbes, or xenobiotics (aromatic hydrocarbons). The involvement of reactive oxygen species is due to their direct action (oxidative gases) or by their intermediates. Therefore, the protective action of antioxidant systems is expected to be found in the lungs, but to a lower extent than in the liver. Most of the defence of the lungs is based on its structure as the early portions of the airways impede the passage of most harmful agents [127]. This makes it more difficult to determine the mechanism of some lung diseases, especially those involving reactive oxygen species.

9.4.1 Mineral Dust

In some parts of the world people from prehistoric times have been exposed to airborne mineral powders and dusts. The main protection the lungs have against this is the ciliary lining of the upper respiratory tract. This is effective against short-term, acute exposure and few develop inflammatory responses in this case. Today, however, many workers undergo chronic exposure to mineral dusts such as asbestos, silica, and coal dust and frequently develop inflammatory reactions as a result. With long-term exposure these works may develop lung diseases such as fibrosis, silicosis, or black lung disease. The mechanism of silicosis was only understood after the importance of inflammatory processes and its consequences were accepted.

Mineral dusts possess large amounts of stable free radicals on their surface, which can absorb iron atoms producing hydroxyl free radicals. Using ESR, the density of these electron donor centers was determined to be approximately 10^{18} per square meter [10, 53, 87, 125]. Anthracite coal has more stable free radicals than bituminous coal and their crushing increases the extent of the active surface. *In vitro*, these particles rapidly produce cytolysis of erythrocytes or other viable cells. The production of free radicals by dust particles proceeds as follows.

Z (activated centers) — O_2 $O_2^{\cdot}$ — $2 H^+$ H_2O_2 — e^- $OH^{\cdot} + OH^-$
$OH^{\cdot}$ (in dust) + LH (phospholipid) $L^{\cdot}$ (peroxide) + H_2O

While it is apparent that smoking and exposure to mining dusts leads to increased free radical production and fibrosis, age is also a contributing factor (see table 9.7). Even in moderate disease, such as pneumoconioses, which occurs earlier than fibrosis, the amount of free radicals in the lungs of miners is high, as is the lipid peroxidation in their blood [125]. However, not all mineral dust exposure leads to fibrosis as polyphosphate covered particles of asbestos or glass fibers do not possess electron donor centers on their surfaces.

Table 9.7 Free radical concentrations in the lungs of coal miners, after Dolal [53].

Age (years)	Smoking (years)	Mining (years)	Disease	Free radical concentration (relative amount)
63±7	40±2.6	29±1.5	Accident	0.42±0.16
64±8	33±2.0	26±1.4	Moderate fibrosis	2.80±1.30
70±8	30±2.6	33±1.6	Massive fibrosis	10.20±2.40
71±6	Nonsmoker	41±5	Massive fibrosis	12.30±5.60
65±10	Nonsmoker	28±3	Lung cancer	3.8±2.3

The sequence of events in the production of fibrosis and silicosis is as follows.

- Inhalation of dust, their reaction with oxygen and adherence to lung epithelium
- Phagocytosis of dust particles by alveolar macrophages and their subsequent activation with the release of reactive oxygen species or free radicals
- Stimulation of the biosynthesis of leukotriene B_4 and necrotic factor, TNF
- Mobilization and activation of neutrophils and eosinophiles, and increased production of IgE and IgG antibodies
- Increased formation of peroxides and aldehydes, which damage DNA and tissues

The PMN leukocytes of miners and fibrotic patients has an increased chemoluminesence emission, which clearly shows they are activated [125, 175]. This sequence of events can only be partly reduced in intensity by administration of the iron chelator desferrioxamine or superoxide dismutase prepared in liposomes [135]. Attempts of the organism to adapt and defend itself are evident in coughing (to expel dust and other irritating material) and increased content of antioxidant systems. The success of these defenses is strongly influenced by age, smoking, and diet [87, 175].

9.4.2 Asthma

As described above, long-term exposure to mineral dust leads to an inflammatory response that is at least partly intended to rid the organism of the irritating agent. This inflammatory response is also involved in other forms of respiratory diseases, such as asthma, allergy, and infections.

Clinically, asthma is characterized by a variable and partly reversible obstruction of the airways due to hyper-reactivity to nonspecific stimulants (allergens, chemicals, cold air). After the acute episode, which lasts about 2 hours, a period of moderate symptoms lasting 6 to 12 hours occurs. Modifications in the airways that are related to an inflammatory response, such as leukocyte accumulation and thickening of the basal membrane, can be seen in asthma.

The inflammatory reaction is triggered by the formation of an allergen-antibody complex that adheres to the membrane of mast cells, epithelium, and macrophages. Next, chemotactic factors are released that attract neutrophils and eosinophiles. Significant amounts of immunoglobulin (IgA and IgG), prostaglandins, and leukotrienes are released by the activated cells. *In vitro*, the leukocytes from asthmatic patients isolated during an acute episode release large amounts of superoxide radicals (15 nmol/10^6 cells). Normal people's leukocytes release much less superoxide radical (7 to 10 nmol/10^6 cells). Also, *in vitro*, the leukocytes from asthmatics release large amounts of hydrogenperoxide (10^{-5} mol/10^6 cells in 2 hours). The increased secretion of hydrogenperoxide and $PGE_{2\alpha}$ are decisive factors for irritation of epithelial cells and hypersecretion of mucus [67, 125, 163].

Alterations of basal membrane induces an influx of calcium ions that activate proteases and phospholipase A_2. The latter releases arachidonic acid from phosphatidylcholine (leaving lysophosphatidylcholine), which is the substrate for the biosynthesis of prostaglandins and leukotrienes. Therefore, a cascade of reactions proceeds producing prostaglandins, lipid peroxides, platelet activating factor, LTB_4, histamine, and serotonin. This cascade of mediators and peroxidative products amplifies the initial episode, triggering the second, long lasting effect. Reactive aldehydes, such as 4-hydroxynonenal and malondialdehyde are also chemotactic factors. They also inhibit adenylate cyclase, glucose-6-phosphatase, and disrupt β-adrenergic receptors [64].

Understanding the basic mechanisms of asthma, including the production of reactive oxygen species, has lead to the development of new drugs to control the condition. These include N-acetyl cysteine, chromoglycate, Ebselen, and phenolic derivatives (E-5110) [91, 168].

9.4.3 Hyperoxia

Chapter 8.4 discussed the modifications arising from hyperbaric conditions and the resulting rise of reactive oxygen species [33, 195]. The exchange of oxygen and carbon dioxide is very efficient, in part, because of the large surface area of the lungs (about 200 m^2). While oxygen is toxic, the organism protects itself at normal oxygen partial pressures with antioxidants. At partial pressures slightly above normal, toxicity can occur and is evidenced by irritation, nausea, and anorexia. At higher partial pressures (over 0.8 atmospheres) pain, dispenia, tracheobronchitis, tusis appears. Clinical manifestations of oxygen toxicity include hyperemia, edema, and swelling of the upper airways. Edema, vasoconstriction and accumulation of fluid in the interstitial spaces are produced by the increased permeability of membranes, in which reactive oxygen species are strongly involved [29, 102].

Neural toxicity, which accompanies this syndrome, is due to dramatic modifications of GABA in the brain. Reactive oxygen species strongly affect the glutamate level and its transport across membranes [45, 111].

Oxygen toxicity is inversely influenced by thyroid and adrenal hormones. For unknown reasons, newborns are more resistant to oxygen than adults. Rats adapted to hyperoxic conditions have an increased content of antioxidant enzymes such as superoxide dismutase and glutathione peroxidase, while those deficient in vitamin E are very sensitive to oxygen toxicity [114, 140, 187].

The complexity of hyperoxia becomes apparent when it is realized the fatty acid content of the lungs influence the biological consequences of oxygen exposure (table 9.8). Only polyunsaturated fatty acids, such as linoleic (C18:2), linolenic (C18:3) and arachidonic (C20:3) acids produce significant modifications. These polyunsaturated fatty acids stimulate the phagocytosis of PMN leukocytes and the subsequent release of reactive oxygen species. At the same time they limit the extension of the immunologic alarm to lymphocyte activation.

Table 9.8 Biological effects of polyunsaturated fatty acids

Stimulation
- C phospholipase from neutrophils
- NADPH oxidase and superoxide production from neutrophils
- ATPases from erythrocytes and endoplasmic reticulum
- Guanylate cyclase from platelets

Inhibition
- Adhesion of lymphocytes
- Adhesion of IgG on lymphocyte surface
- Incorporation of amino acids into brain

Fatty acids circulate in the blood in a concentration that ranges between 300 and 500 µM, but most are bound to albumin. The free amount is only 1% of the total. In some pathological conditions, such as severe stress, diabetes, intense physical effort, ischemia, or traumatic shock, the fatty acid content of the blood increases, especially the free fraction. In addition, fatty acid incorporation into cells increases. The increase of polyunsaturated fatty acids in the cytoplasm offers conditions suitable for peroxidation. The aldehydes resulting from peroxide decomposition are released into the extracellular space and the plasma. Pulmonary edema and respiratory failure can involve the direct involvement of polyunsaturated fatty acids or the indirect effects of prostaglandins PGE_2 and $PGF_{2\alpha}$. This is demonstrated by the infusion of polyunsaturated fatty acids into the lungs, which leads to capillary vasoconstriction, increased permeability, and a strong release of histamine. These modifications may be blocked by the previous administration of vitamin E and promethazine.

This experiment may partially explain the observed resistance of newborns and some animals to hyperoxia [153, 173, 204]. Animals with an increased content of polyunsaturated fatty acids are more resistant to peroxidation because they have a rich supply of antioxidants. At the same time, there is an increase in extracellular superoxide dismutase activity that supports the intracellular defense [151, 168, 199, 202]. Similar results were obtained with the administration of chemical compounds such as paraquat, bliomycin, naphthylthiourea, or experimentally induced ischemia. In all these conditions, high levels of reactive oxygen species were produced [31, 83, 108, 116].

9.4.4 Smoking

The harmful effects of smoking have been well documented. Reactive oxygen species are strongly involved in the effects of smoking, as each puff of cigarette smoke contains 10^{14} free radicals in the gas and 10^{15} free radicals in the tar. Free radicals are also found in cigarette ash. The signals obtained from ESR studies of the ash are similar to those produced by reactive phenols [100].

Cigarette smoke and tar also contain large amounts of polyaromatic hydrocarbons, most of which have carcinogenic properties. Other carcinogens in cigarette smoke include toxic metals such as lead and cadmium [127].

The lungs react to exposure to smoke by mobilizing phagocytic leukocytes. Therefore, all smokers develop a persistent inflammatory response, which involves the release of reactive oxygen species. The magnitude of this response depends on the person's age, the number of cigarettes smoked, and individual resistance. Hulea and Olinescu have shown that oxidative stress in smokers is apparent in the increased levels of leukocyte activation, plasma lipid peroxidation, and total antioxidant and glutathione reduction. It is interesting to note that the body makes an attempt at adaptation by increasing the amount of ceruloplasmin, fibrinogen, and glutathione peroxidase.

Considerable scientific literature shows that tobacco smoke is a causative or aggravating factor in the etiology of cardiovascular disease or lung cancer. The following discussion may explain the consequences of oxidative stress in smokers (figure 9.3). The presence of free radicals in cigarette smoke both directly and indirectly reduces the antioxidant content of the organism. The free radicals directly react with antioxidants and also activate leukocytes and macrophages. The activated cells produce reactive oxygen species that also react with antioxidants. At the same time there is a compensatory increase in ceruloplasmin, fibrinogen, and antioxidant enzymes in the erythrocytes. This provides some additional protection, but the long-term net effect is increased lipid peroxidation. These factors all contribute to oxidative stress, which leads to increased LDL oxidation, platelet activation, and vascular endothelium injury. The vascular insult proceeds along two possible pathways. One involves the formation of foam cells and plaque, and the other involves platelet-fibrinogen interaction and thrombosis.

9.4.5 Adult Respiratory Distress Syndrome

Adult respiratory distress syndrome (ARDS) is a devastating disease that is characterized by an edemous reaction in the lungs that leads to a defective gas exchange. People with ARDS frequently develop lung trauma and infection [143]. The mobilization of PMN leukocytes in these patients releases peroxides, nitric oxide, TNF, or other cytokines. The blood of these patients contain elevated lipid peroxides and decreased protein thiols and linoleic acid [100, 143, 145].

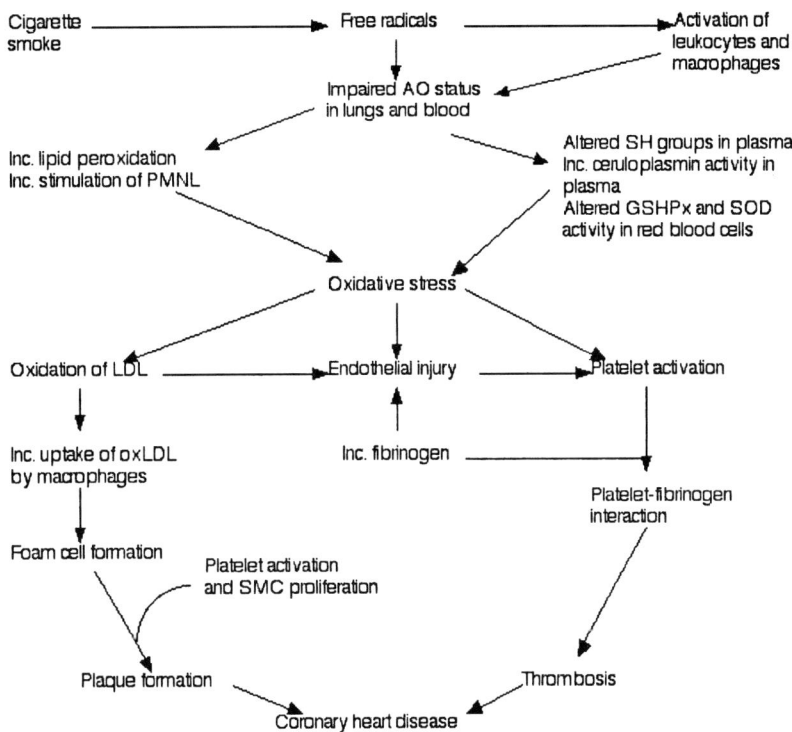

Figure 9.3. Probable scheme of oxidative stress in smoking leading to heart disease.

9.5 LIVER FAILURE

The liver is the key organ in antioxidant defenses. The liver contains nearly all physiological antioxidants in the highest amount of the body. This organ is the main site for detoxification of xenobiotics and the decomposition of surplus reactive oxygen species or free radicals. This is the reason the highest levels of lipid peroxides in plasma are found in cases of severe liver failure [124, 167, 203]. Therefore, it is to be expected that when chemical intoxication (involving reactive oxygen species) occurs, the increased formation of peroxides and decrease of antioxidants will only occur when the liver's pool of antioxidants are exhausted. It has been demonstrated *in vitro* with liver microsomes that lipid peroxide formation starts only after intracellular glutathione drops to a very low level [6, 27, 35, 48, 60, 90].

The chemicals that damage the liver may be divided into three groups.

I. Compounds like carbon tetrachloride, bromoform, iodoform, vinyl chloride, and halatene. These compounds easily form free radicals (such as the CCl· radical from carbon tetrachloride) that readily react with glutathione, produce lipid peroxides, or bind covalently with DNA or other macromolecules [157, 178].

II. Compounds like diethylmaleate, di- and trichloroethylene, chloroform, allylic alcohol, or acrylamide, all of which are chemical pollutants. These compounds or their electrophilic metabolites act directly with glutathione producing lipid peroxides and decreases levels of glutathione. These compounds are metabolized in the endoplasmic reticulum [6, 27, 90].

III. Compounds formed by redox reactions such as cytostatic drugs, hydrozine derivatives, nitrofurantoin or alloxan, or 6-hydroxydopamine. These compounds are metabolized into semi-quinonic free radicals that react with glutathione. Hydrogenperoxide is formed during the reaction and lipid peroxides are produced [108, 178, 203].

Experiments involving chemical intoxication of the liver demonstrate the existence of a long incubation period, during which the reactive intermediates (mostly free radicals) react with glutathione and other antioxidants. During this incubation period, which varies depending on the extent of exposure and individual resistance, oxidation occurs, but it is directed at eliminating the antioxidant systems. An incubation period appears in all chronic liver intoxications, including that caused by alcohol. Towards the end of the incubation period, nearly all the principle antioxidants are exhausted and lipid infiltration and peroxidation of the liver rises rapidly [126]. Protection from these effects is provided by desferrioxamine (iron chelator) and calcium blocking agents (Nifedipin, Verapanyl, Diltyazem), all of which exhibit antioxidant properties [135, 190].

The chemical intoxication of liver always leads to increased formation of lipid peroxidation in the blood plasma. Liver failure caused by viral or bacterial infections produce similar results. A possible mechanism has been proposed by Halliwell and Guttteridge [71, 72]. This mechanism implicates iron ions as an alternative to the bilirubin catalyzed reactions described in section 5.3.3 [126, 131].

The implications of free radicals and the biochemistry of alcohol were stated by Di Luzio who, in 1963, claimed that acute ethanol-induced fatty liver was prevented from developing by antioxidant. In 1967 he claimed that increased lipid peroxidation is also produced. After twenty years, these statements are hotly debated. In 1972, the English biochemist T. F. Slater dedicated an entire chapter to the hepatotoxic effects of alcohol in his book, *Free Radical Mechanisms in Tissue Injury*. With the introduction of new techniques in free radical research (ESR, chemiluminescence, alkone determinations) new insight has been found [177, 178]. A recent book [127] states that the toxic effects of alcohol are cumulative with chronic exposure, and, even under experimental conditions, the production of fatty liver requires a minimum of 4 months incubation. Under normal use conditions by humans five to twenty years of exposure are required [68, 126]. In general, the implication of peroxidation as a factor in liver damage from chemical or microbial effects is well demonstrated [1, 27, 35, 48, 126].

During sustained, chronic alcohol intake, the nonspecific antioxidant defence is continuously challenged in order to cope with the increased formation of free hydroxyethyl radicals ($CH_3C^{\cdot}HOH$, $C^{\cdot}H_2CH_2OH$), hydroxyl radicals, and other reactive oxygen species as well as lipid peroxides and the associated aldehydes [127, 178]. The increased production of hydrogenperoxide due to increased peroxisome and mitochondrial activity leads to another wave of reactive oxygen species. The metabolism of alcohol forms the end product acetaldehyde, which is normally metabolized by aldehyde dehydrogenase. Excess acetaldehyde is metabolized by xanthine oxidase, which adds another source of hydrogenperoxide. The continuous and increased formation of reactive oxygen species leads to a decrease in antioxidants, especially of vitamin C and E, which may be exacerbated by malnutrition present in many heavy drinkers.

9.6 PHYSIOLOGICAL BRAIN DISEASES

Among the organs of the body, the brain is unique from all points of view, including that of free radicals and reactive oxygen species. The brain is poorly endowed with antioxidant protective systems, so its main line of defense is the blood-brain barrier. This barrier blocks the penetration from the blood into the brain of many high molecular weight compounds. Lipophilic compounds (alcohol, drugs like phenothiazine, barbiturates) and metallic ions easily penetrate into the brain where they have various biological, biochemical, or psychic properties.

The principle antioxidants found in the brain are copper-zinc superoxide dismutase, melatonin, and ascorbic acid. The brain is rich in oxidative substrates, such as the polyunsaturated fatty acids. Due to the brain's particular properties, the direct implication of reactive oxygen species is difficult to obtain in humans. The evidence of an effect of reactive oxygen species on the brain of humans in almost totally extrapolated from animal studies.

The brain is highly vulnerable to attack by reactive oxygen species because it does not have an adequate antioxidant defense system. Because of its high metabolic activity, the brain uses large amounts of molecular oxygen. In a conscious, young individual oxygen is used at the rate of 35 ml/100 g neural tissue per minute. This means that the brain, which accounts for only 2% of the body's weight, uses 20% of the total inspired oxygen when the individual is at rest. In the brain, oxygen mainly participates in the oxidation of carbohydrates and generates approximately 4×10^{21} molecules of ATP per minute. Because the use of oxygen in the brain is high, it follows that the generation of reactive oxygen species may also be high. The high use of oxygen is not the only reason for a high risk of reactive oxygen species attack. The brain is also enriched in polyunsaturated fatty acids. Cerebrospinal fluid contains little ceruloplasmin, transferrin, or other antioxidants, but does contain low molecular weight complexes of iron and copper [75].

Neurotransmitters, which are abundant in some zones of the brain, generate massive amounts of reactive oxygen species. The glutamate receptor is known to be involved in neurological diseases. Therefore, the damage caused to glutamate by reactive oxygen species will be concentrated in the neural structures that contain the highest amounts of this neurotransmitter [79, 118]. With longevity, the damage related to glutamate release accumulates, leading to the gradual morphological and physiological destruction of neurons that are exposed to this transmitter [118, 174]. A possible sequence of events in this is shown in figure 9.4.

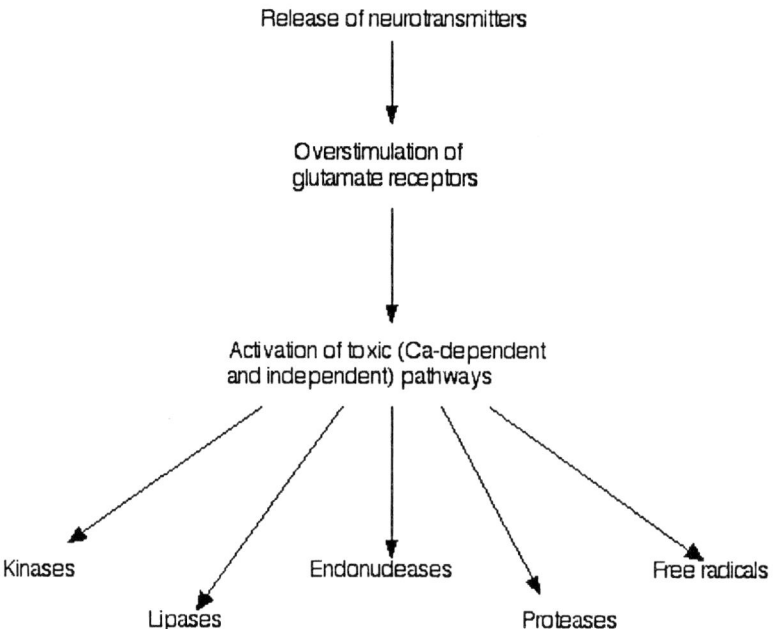

Figure 9.4. Possible formation of free radicals in cerebral disorders through the release of excitatory amino acids (glutamate).

- The activation of dopamine-β-hydroxylase, which catalyzes the transformation of dopamine into noradrenaline is associated with hydrogenperoxide formation [34, 79].
- During melanin biosynthesis the action of tyrosine and peroxidase favor reactive oxygen species production [75].
- Metabolism of catecholamines releases hydrogenperoxide [21, 34].
- Oxidative metabolism of catecholamine does not normally occur physiologically, but may arise when catecholamines are released in high amounts (such as under emotional stress) [21]. Usually, catecholamines, once released from neural synapses, are metabolized by MAO or COMT reactions. With high amounts of catecholamines, oxidative reactions may occur, leading to the formation of quinonic forms (adrenochrome, dopachrome). These derivatives possess hallucinogenic, mutagenic, and antitumor properties. The low frequency of cancer among patients with mental diseases has been suggested to be due to the increased formation of adrenochromes [21, 65, 79].
- The complex events of aging take place mainly in the brain, where granules of lipofuscin pigments (which include peroxides) are formed [77].
- In addition to physiological processes, physical trauma, bacterial infection, or hemorrhage increase the potential for the production of reactive oxygen species due to inflammatory processes [32, 79, 176].

Therefore, the physiological conditions in the brain are amenable to the formation of reactive oxygen species. However, the low levels of antioxidants present are able to mostly keep this under control. If additional sources of reactive oxygen species are introduced, damage may occur. This damage requires a long time to develop and the rate is dependent on individual resistance [65, 75].

In table 9.9, we present some clinical conditions where significant levels of reactive oxygen species and peroxides were detected in the blood or the brain was found to have greatly increased stores of lipofuscin. Most of these conditions have a reversible stage in which no significant levels of peroxidation were detected [75, 77].

Table 9.9 Clinical conditions in which significant reactive oxygen species were detected in the brain, after Cross [39].

Hyperbaric oxygen
Neurotoxin intoxication
Chronic intoxication with metals (Al, Mn, Pb)
Cerebral trauma, hemorrhage
Cerebral ischemia
Ataxic telangiectasia
Allergic encephalitis
Demylating diseases
Alzheimer disease
Parkinson disease

Additional sources of reactive oxygen species in the brain may arise in some pathological conditions such as cerebral ischemia [79, 174]. Only five minutes of ischemia is sufficient to produce significant amounts of hydroxyl free radicals as is demonstrated by the increased salicylate hydroxylation and loss of glutamine synthase. Experimental addition of superoxide dismutase and catalase limit the neurological lesions produced by ischemia.

New evidence of the involvement of reactive oxygen species in pathological conditions include [2, 118, 199]:

- 6-hydroxy dopamine formation as a qualitative modification of dopamine metabolism. This compound is a powerful oxidant [79].
- Elevated levels of calcium enhance free radical induced damage to cerebral morphological lesions and calcium blockers may offer partial protection [190].
- Activation of the NMDA (N-methyl-D-aspartate) receptor site has been implicated in the post-ischemic elevation of lipid peroxidation in the hippocampus. NMDA agonists are especially potent in the stimulation of superoxide to form the intensely oxidative nitro peroxyl radical. There is evidence that nitric oxide mediates the neurotoxicity of glutamate [145].

Excitatory events may stimulate reactive oxygen species release, but there is also evidence for a reciprocal relation between these phenomena. Reactive oxygen species can lead to extracellular release of glutamate or can release mitochondrial calcium into cytosol. Therefore, a reciprocal relation between excess neural activity and excessive reactive oxygen species formation seems likely.

As Alzheimer disease occurs with increasing frequency, so does the research about the possible implication of reactive oxygen species. Various exogenous agents have been claimed to be involved in the etiology of Alzheimer disease, including granules of alumino silicate, amyloid in copper complexes, and neurotoxic endogenous compounds arising from qualitative derivation of normal metabolism of dopamine or serotonin. Such derivatives include adrenochrome or tryptamine-4,5-dione (4,5-DKT). The later compound acts in the limbic system of the brain by producing hydrogenperoxide, oxidizing sulphydryl groups of proteins, and covalently binding to nucleotides. It also produced amyloid microangiopathy, which affects cerebral capillary blood circulation [32, 174].

Finally, we should mention that the antioxidant system in the brain decreases with aging. Superoxide dismutase activity especially decreases in the frontal cortex, septum, caudal nuclei, and substantia nigra, all of which are rich in catecholamine releasing neurons. As for other reactive oxygen species associated with diseases, a diet rich in antioxidants should offer partial protection [65]. Unlike other conditions involving reactive oxygen species, oxidative stress in the brain is more difficult to repair and is more difficult to treat because of the blood-brain barrier.

9.7 OTHERS

9.7.1 Diabetes

Although there have been reports of increased levels of reactive oxygen species in the blood of diabetics [16, 161], the implications of free radicals in the etiology of the disease is still unclear. Diabetes may be produce experimentally by the administration of alloxan (a purine derivative) and streptozotcyn (a bacterial toxin), which results in the formation of large amounts of reactive oxygen species.

In both experimentally induced diabetes and in most patients, increased levels of peroxides are detected. The most increase in peroxide content is associated with the most severe forms of the disease, especially when they are connected with cardiovascular disease or retinopathy [133, 172].

It should be noted that the pancreas is one of the poorest organs in content of antioxidants, and the supplementation of antioxidant vitamins provide protection for animals against experimentally induced diabetes [161].

In addition to individual sensitivity (influenced by heredity), obesity also plays an important role in the etiology of type 2 diabetes (adult onset or insulin independent). Fat deposition within skeletal muscle may be a characteristic leading to insulin resistance by disturbing the oxidative capacity of the muscle.

Experimentally induced diabetes develops when alloxan is converted to dialuric acid and by autooxidation to reactive oxygen species, especially hydroxyl. Alloxan selectively destroys pancreatic beta cells, inducing type 1 diabetes (insulin dependent). The addition of exogenous scavengers that are specific for hydroxyl radical, such as thiourea or superoxide dismutase and catalase, reduce the extent of the drugs effect [133, 161].

The unresolved problem is the source of the increased lipid peroxidation. It is not clear if it is a consequence of disturbed lipid metabolism and oxidative stress or if it is a consequence of increased plasma glucose levels. Glucose at high concentrations under physiological conditions may undergo autooxidation, becoming a source of free radicals. In severe diabetes, a significant increase in the production of glycosylamines occurs. Glycosylamine formation involves the nonenzymatic reaction between protein and glucose. It occurs in many tissues under diabetic conditions, especially the liver, kidney, and retina [16, 161]. This process involves the interaction of a free amino group (lysine) with reducing sugars, hydroxyaldehydes, or ketones. The modified proteins undergo a Schiff base rearrangement to form more stable glycosylamines called Amadori compounds. The formation of these compounds includes the intermediate appearance of free radicals [16]. During the development of diabetic atherosclerosis, glycosylamine formation and lipid peroxidation can occur at the same time as LDL modification by autooxidized glucose.

9.7.2 Cystic Fibrosis

Cystic fibrosis is a highly inflammatory, congenital disease involving an increased leukocyte mobilization and activation. The leukocytes of these patients shows a significantly increase in chemiluminescence [125, 143].

Lung damage results from ion transport abnormalities in the airway epithelium leading to mucus accumulation and bacterial invasion. Highly reactive oxygen species released by activated leukocytes are partly responsible for microvascular and tissue injury. Patients with cystic fibrosis often develop diabetes and have a short life expectancy [33, 67, 81].

Among the epithelial compositional alterations in cystic fibrosis is a decrease in the concentration of polyunsaturated fatty acids (18:2 and 20:4) and a decrease in the amount of vitamin E, selenium, and glutathione. Higher levels of lipid peroxides were found in the blood of cystic fibrosis patients [78, 131, 146].

9.7.3 Ulcer

A common treatment for inflammatory conditions is the use of nonsteroidal antiinflammatory drugs (NSAIDs) such as indomethacin or aspirin. A side effect if extensive use of these drugs is irritation to the stomach lining that can lead to a gastric ulcer. NSAIDs are used because they inhibit cyclooxygenase, which is involved in the synthesis prostaglandins. Prostaglandins synthesized in the gastric mucosa have a vasodilatory action, inhibit gastric acid secretion and are now used in the treatment of gastric ulcer.

Reactive oxygen species may also be involved in some intestinal problems. Bacteria in the intestine may lead to inflammation, which results in the release of reactive oxygen species. Increased reactive oxygen species formation leads to increased permeability of the intestinal mucosa, allowing infiltration of leukocytes [133, 194].

Chapter 10

THERAPEUTIC DRUGS AND NUTRITION

10.1 INTRODUCTION

In this chapter we will deal with the interesting and controversial question of the relationship between free radicals and antioxidants on the one side and the treatment of disease, especially alternative medicine, on the other.

As has been described in the book up to this point, living beings that require an oxygen containing atmosphere are also exposed to the toxic effects (oxidative stress) of oxygen. In addition, other endogenous and exogenous factors contribute to the overall oxidative stress experienced by the organism. The second point is that the organism must maintain an antioxidant system to protect vital cell structure and function from the effects of oxidative stress. The third point is that this antioxidant system functions largely by becoming the preferred target of reactive oxygen species. This means the components of the antioxidant system must be constantly replaced. Some, such as the antioxidant vitamins must be obtained from the diet.

These three conclusions lead to two problems.

- While the advertising and health food industries strongly promote the benefits of antioxidant supplements, most medically trained people are unenthusiastic or even skeptical about their benefit. In part this is because of the complexity of the free radical and antioxidant reactions, which are difficult even for biochemists to understand. Another factor is that the research has produced results about the benefit of antioxidants that are far from clear cut. It does not help that those who promote the use of antioxidants make claims that they are able to cure every conceivable ailment, from corns to cancer.
- Secondly, the message about antioxidants is subtle, which makes the message easily misunderstood. If there is any benefit of the use of antioxidants, the benefit is one of protection, not a cure. In addition, the protection is not absolute. The degree of protection depends on many other factors including inherited factors of resistance, physical and emotional stress, and the presence of other diseases that cause inflammation, etc. It is quite possible for even the antioxidant levels provided by a

supplemented diet to be overwhelmed by the load of free radicals and reactive oxygen species.

To further complicate the understanding of the role of antioxidants and free radicals is the point that is not well understood by non-scientists, that free radicals are continuously produced in the body during its normal functioning. It should also be clear from the evidence presented earlier in this book, that focusing on a single antioxidant as being some sort of magic potion against free radicals is foolish. There are many compounds that have antioxidant activity and all have unique roles.

Another fact that is not clearly understood, especially by lay people, is that the harmful effects of oxidative stress are reversible when they occur at low levels. Only when the damage exceeds a certain point is it detectable as signs and symptoms and only at some greater point is the damage permanent. Furthermore, long-term exposure is generally necessary to deplete the antioxidant systems to the point where permanent injury can occur.

The marketplace is flooded with antioxidant or other health products that are mostly extracts from exotic plants for which it is claimed they are an universal drug. This situation was made worse in 1994 when the United States Congress passed the Dietary Supplement Health and Education Act. With this law, vitamins, minerals, and herbs were classified as food supplements rather than drugs. This reduces the control the Food and Drug Administration is able to exert on them. Prior to this the manufacturers of these products could not make explicit healing claims. Now they openly do so. These herbal extracts do indeed contain some biologically active compounds having various effects. However, as pointed out in the May 1996 issue of Newsweek, frequent side effects and even death have been recorded after ingesting some plant extracts.

10.2 PROOXIDANT DRUGS

The main free radicals arising in biological media were described in chapter 2 as quinones and aromatic compounds. Many drugs also contain this structure. Most drugs that have a cytostatic action are effective because of the free radicals generated as the drug goes through a redox cycle.

The list of prooxidative drugs is not conclusive because of the few studies that have been done on the question. Even for those drugs mentioned in table 10.1, their prooxidative action was only demonstrated during experiments to determine their mode of action.

Quinones (menadione, cytostatics) are metabolized through the action of cytochrome P_{450} dependent enzymes in the endoplasmic reticulum and produce large amounts of free radicals. These reactive oxygen species react with glutathione and other antioxidants as well as with hemoglobin (menadione). Increased lipid peroxidation and activation of calcium-dependent endonuclease also occur. In spite of differences in the specific action, all these drugs have side effects such as hemolysis (menadione), cardiomyopathies (cytostatics), and liver injury (paracetamol). Some antioxidants, such as N-acetyl cysteine (NAC), or calcium channel blockers may reverse these side effects.

Paraquat poisoning attacks the lungs. Anthracycline cytostatics attack the heart. The remainder of the compounds mentioned in table 10.1 induce hepatotoxicity upon either chronic or acute exposure.

Table 10.1 Drugs and other compounds that produce free radicals when metabolized.

Compound	Type of Free Radical	Use
Vitamin K_3	Semiquinone	Anticoagulant
Adriamycin Daunomycin	Anthracycline	Cytostatic
Mitomycin Streptonigrin	Semiquinone	Cytostatic
Acetaminophen	ROS*	Antipyretic Analgesic
Metronidazole	ROS	Bactericide
Nitrofurantoin	Semiquinone and ROS	Bactericide
Paraquat	ROS	Herbicide
Phenyl hydrazine and its derivatives	ROS	Drug industry
Organic solvents (chloroform, carbon tetrachloride, etc.)	Free radical based on the structure of the solvent	Chemical industry

*Reactive oxygen species

All these compounds damage glutathione, so maintaining an adequate level of this antioxidant is essential for cell survival. Unfortunately, compounds that release free sulphydryl groups, such as NAC or cystamine, are likely to produce side effects such as gastric ulcer or hypotension.

Several compounds, such as ascorbic acid, menadione, phenylhydrazine, or nitrites, exhibit a paradoxical behavior *in vitro*. Their hemoglobin binding capacity is significantly increased by purified superoxide dismutase when it is added to a suspension of erythrocytes. The mechanism involves the following reactions.

Redox compound + O_2 Intermediate + $O_2^{\bullet}$ (superoxide)
Hemoglobin + $O_2^{\bullet}$ + 2 H^+ Met-hemoglobin + H_2O_2
2 $O_2^{\bullet}$ + 2 H^+ — superoxide dismutase O_2 + H_2O_2

In the presence of purified superoxide dismutase, an increased amount of hydrogenperoxide is formed, which increases the oxidative stress [2, 199]. This emphasizes the necessity of including purified catalase with superoxide dismutase, to decompose the hydrogenperoxide.

Unexpected modifications may arise following the experimental administration of toxic solvents. The concomitant administration of zinc salts protects against solvent intoxication. The explanation includes the increase of glutathione and metallothionine after zinc administration, which protects against reactive oxygen species formation by the organic solvent [138].

10.3 ANTIOXIDANT DRUGS

Several drugs with antioxidant action, including flavonoids and NSAIDs, were discussed in chapter 7. Some drugs that act on the central nervous system, such as phenothiozines and barbiturates, are also effective antioxidants. Given that iron has a prominent role in oxygen activation and the formation of reactive oxygen species, the chelator desferrioxamine might be considered a drug with antioxidant properties. In practice, desferrioxamine is used to control iron overload following the use of cytostatic drugs or in the case of diseases such as hemochromatosis. Therefore, if the role of free iron ions is as important as some scientists think [71-74], the administration of desferrioxamine should be effective against tissue damage due to reactive oxygen species. In fact, as shown in table 10.2, the action of this drug is much broader. However, the therapeutic use of the drug is limited [135].

Table 10.2 Antioxidant effects of desferrioxamine. After Cross [39].

---

1. Significantly increases the actived period of stimulated neutrophils.
2. Decreases formation of pulmonary lesions induced by activated neutrophils and complement.
3. Antiinflammatory action in high doses for experimental models.
4. Blocks the hemolytic action of some chemicals.
5. Inhibits the development of autoimmune diseases in experimental animals.
6. Decreases hydrogenperoxide toxicity towards isolated hepatocytes.
7. Reduces paraquat toxicity in rats and mice.
8. Decreases alloxan toxicity in the formation of experimental diabetes.
9. Decreases carbon tetrachloride toxicity in experimental animals.
10. Potentiates treatment in some neural degenerative diseases.
11. Beneficial effects on patients with iron overload.
12. Beneficial effects in heart or kidney ischemia or skin transplant.
13. Beneficial effects in children with b-thallasemia major.
14. Beneficial effects in human leukemia and neuroblastoma.
15. Beneficial effects in patients undergoing kidney dialysis and intoxicated with aluminum.

Desferrioxamine is a trihydroxamic acid derived from a plant source. It has a high affinity for iron (III) and therefore inhibits the Haber-Weiss reaction. The drug is manufactured by Geigy Company. It may be injected sub-cutaneously at doses up to 60 mg/kg body weight to prevent its concentration exceeding the threshold of 1 mM. It may also be administered orally and is rapidly excreted by the kidneys and in the bile.

In spite of an abundance of studies regarding the biological properties of iron chelators, no clear conclusions on their usefulness has been obtained. Complexes of iron with chelators such as EDTA, DTPA, and DETAPAC stimulates lipid peroxidation in low concentrations (5 µM) and inhibits peroxidation at higher concentrations (60 µM). Natural iron complexers, such as transferrin or lactoferrin are excellent antioxidants when they contain a low iron load, but become prooxidant when they have a full iron load. New therapeutic chelators of iron are being tested (rodotoulic acid, phenathroline, 2,3-dihydroxibenzoate, pyridoxal hydrazone).

Disulfiram (Antabol, Antabuse) is a drug used to treat chronic alcohol abuse. Once ingested it is metabolized to diethyldithiocarbamate (DDC), which contains a free sulphydryl group. DDC inhibits aldehyde dehydrogenase, so the metabolism of alcohol is stopped at the level of acetaldehyde. Acetaldehyde is moderately toxic is moderately toxic and produces unpleasant symptoms such as nausea. In fact, DDC inhibits many enzymes, including dopamine-b-hydroxylase, hexokinase, xanthine oxidase, and superoxide dismutase. The last two of these enzymes have opposing actions in the formation of reactive oxygen species. During chronic alcohol intoxication, an increased free radical formation takes place, but raised levels of lipid peroxides only appear when liver failure occurs [68].

The administration of disulfiram to humans can result in toxic effects when other drugs, such as antipyrine or coumarin, are taken concomitantly. This results in interference with the MEOS system, associated with cytochrome P_{450} in the liver. On the other hand, DDC has a great capacity to bind free copper, and it is effective in treating acute intoxication with paracetamol, benzoquinone, or bromobenzene. DDC forms a complex with quinones preventing the formation of semi-quinone free radicals. Therefore, DDC provides a good protection against chemicals that damage the liver, but stimulates the harmful action of ozone and paraquat on the lungs. DDC is an antioxidant as it scavenges several free radicals and exhibits moderate radioprotection.

Ebselen is an antioxidant drug produced by Professor H. Sies of the University of Dusseldorf, Germany. Its chemical structure (2-phenyl-1,2-benzo-seleno-3-one) is an aromatic, selenium-containing drug. It is designed to mimic glutathione peroxidase activity. *In vitro* experiments show the drug has a remarkable antioxidant activity and lacks toxicity. However, *in vivo* experiments have not been so positive.

While iron complexing agents possess complex structures that do not allow their penetration of membranes (except DDC), other antioxidant drugs have an aromatic ring similar to the polyphenols (like resorcin). Thus, several synthetic antioxidants have structures that are similar to amino acids. BHT, BHA, or aspirin have a similarity to tyrosine. Ebselen has a similarity to tryptophan, and DDC has a similarity to methionine. Phenylbutazone has a pyrrol ring and two aromatic rings. Taking into account the great reactivity of amino acids toward oxidants, or their special position at the catalytic site of some enzymes, it is possible that several synthetic antioxidants will function to protect these amino acids or compounds like them.

Allopurinol is a specific inhibitor of xanthine and an effective antioxidant. It is used therapeutically to limit the production of uric acid in gout. Its antioxidant activity was successfully used to limit damage resulting from ischemia and hemorrhagic shock. These useful clinical applications are also based on the property of allopurinol to protect the nucleotide pool (ATP) against enzymatic degradation and, consequently, to favor recovery from these accidents [66].

Several studies have shown that most of the calcium channel blockers are also effective antioxidants. Their antioxidant efficiency decreases in the order lacidipine > nicardipine > verapamil > probucol. These calcium antagonists are most efficient against lipid peroxidation. As calcium decompartmentalization and increased permeability of membranes are major consequences of lipid peroxidation, the antioxidant action is related to opposing this effect to limit tissue damage [57, 74, 97].

Another interesting aspect of both drug and diet influences on antioxidant defense is the release of polyunsaturated fatty acids from cell membranes. A presented in chapter 4, this is a nonspecific event that takes place during membrane damage. Thus, by maintaining an intact cell membrane, it is possible to limit or stop the evolution of oxidative stress. Indeed, some drugs seem to act according to this mechanism. These drugs include phenothiosines and procaine, which mimic the action of prostaglandin and vitamin E.

The event catalyzed by phospholipase is the key step for prostaglandin synthesis. At the level of the cell membrane, various intercellular signals are receive (chemical messengers) that trigger intracellular responses. This leads to the activation of phospholipases and the release of arachidonic acid. Thus, phospholipase A_2 is stimulated by calmodulin and variations in calcium concentration, and inhibited by phenothiazines, quinacrine, and corticosteroids. Phospholipase C acts on phosphoinositol and is stimulated by muscarinic agonists.

10.4 LIPIDS

Polyunsaturated fatty acids are an excellent substrate for reaction with reactive oxygen species. Because of this, most antioxidants act at the level or location of these fatty acids. Some of these polyunsaturated fatty acids are required by the body, but cannot be synthesized by mammals. These essential fatty acids must be obtained through the diet.

Essential fatty acids have two classes: the ω3 and ω6 essential fatty acids. Omega 3 fatty acids are found mostly in plant sources and ocean creatures while omega 6 fatty acids are found in animals. Lipids are required for the formation of cell membranes. Essential fatty acids are involved in the formation of prostaglandins, thromboxanes, and leukotrienes, which are autocoids.

Fat in the diet has been targeted in all industrialized countries as a major factor in causing certain diseases, such as cancer, cardiovascular disease, obesity, and stroke. Through a period of centuries, people have eaten a high fat diet because they needed the large amount of calories provided by lipids. Only as manual labor has decreased has a diet high in fat meant an excess of calories was available. In the late 1960s and the 1970s a major campaign was begun in the United States to blame fats for these diseases. In the 70's saturated fats were blamed for causing heart disease. In part this was due to the observation that atherosclerotic plaque has a high content of lipid, oxidized lipid, and oxidized cholesterol. This made foods high in saturated fats (red meat) undesirable and foods high in unsaturated fats (fish and plant sources) desirable.

In the 1980s this simplistic view broke down. Palm and copra oils were found to be rich sources of saturated fats and lean pork was found to be an excellent source of vitamins and minerals. Large variations, based on geographic area, were found in the frequency of cardiovascular disease, diabetes, cancer, etc. that might be correlated with diet [24]. In addition, several studies showed that LDL cholesterol is easily oxidized and then becomes involved in atherosclerotic plaques while HDL cholesterol behaves like an antioxidant [11, 54, 59, 100, 165, 203]. The connection between oxidized LDL and atherosclerotic plaques was found to be macrophages, which are converted to foam cells when they ingest oxidized LDL and trigger the endothelium and platelets to release more reactive oxygen species [11, 129, 152, 196].

Research in other directions found that polyunsaturated fatty acids are excellent substrates for the action of reactive oxygen species. However, not all polyunsaturated fatty acids are equally sensitive. The differences in sensitivity are related to the number of unsaturated bonds in the molecule as well as its geometric organization.

Epidemiological studies showed some lipids are desirable, especially the $\omega 3$ fatty acids [64, 85, 98, 153, 173, 202]. Table 10.3 presents typical results of such studies. The correlation between lower cardiovascular disease and increased essential fatty acids was noted. Greenland eskimos and the Japanese eat a diet much higher in ocean-derived foods than do Europeans or Americans. A diet high in polyunsaturated fatty acids from fish and sea food were claimed to prevent cardiovascular disease, rheumatism, colitis, and auto immune diseases (lupus) [24, 98, 153, 173]. However, practical problems in reproducing the diet in a large population and questions about the extent of the benefit have prevented widespread adoption of such a diet. For example, to gain the maximum benefit, the sea foods must be eaten raw, as cooking decreases the polyunsaturated fatty acid content [173, 202].

Additional studies that were done failed to support the simplistic version of the reduced fat diet being recommended. For example, scientists at the United States Department of Agriculture found that limiting fat intake to 30% of total calories will not necessarily lower LDL cholesterol if the fat consumed is mostly saturated. A diet based on combinations of lauric and myristic fatty acids resulted in higher serum cholesterol than did a diet rich in palmitic acid in healthy normocholesterol young men.

Hassan [163] analyzed oils for their polyunsaturated fatty acid content and fed mixtures to rats. He then determined the amount of lipid peroxidation that occurred in the liver of the fed rats (table 10.4). The results are difficult to interpret. It is clear that more peroxidation occurred in the rats fed large amounts of polyunsaturated oils, but there was also an increase in peroxidation in the rats fed saturated oils. Saturated fats are poor substrates for peroxidation. If peroxidation is harmful, as has been demonstrated in many situations, than the diet high in polyunsaturated fats was less desirable than the diet high in saturated fat. This contradicts the wisdom that unsaturated fats are healthier than saturated fats.

Table 10.3 The frequency of mortality from cardiovascular disease and its possible relationship with the polyunsaturated fatty acid (PUFA) content of the diet [64, 98, 173].

Country	Platelet PUFA content (%)	Arachidonic/ Linoleic acids	w3/w6	Mortality (% of total)
Europe, USA	20 - 26	0.1 - 0.7	50	40
Japan	18 - 22	1.0 - 2.5	12	12
Greenland eskimoes	8.3 - 9.0	6.4 - 8.0	1.2	7

Table 10.4 Amount of lipid peroxide (as TBARS*) and rate of peroxidation in rat liver microsomes fed food oils. After Hassan [78].

Type of oil	Lipid peroxides (nmoles/mg)	Rate of peroxidation (nmoles/mg/min)
Non lipid diet	1.82±0.30	0.60±0.08
Coconut oil	2.60±0.20	1.22±0.17
Lard	2.82±0.50	1.72±0.30
Corn oil	13.60±3.50	1.42±0.15
Herring oil	20.80±4.60	3.73±0.34
Mixed oils	8.76±0.80	1.98±0.09

*Thiobarbituric reactive substances

On the other hand, it has been shown that the toxic effects of ethionine (a hepatocarcinogen involving lipid peroxidation and the subsequent release of transaminase) can be neutralized by feeding rats a diet rich in choline (required for the synthesis of phospholipids) [157, 173, 202].

Other studies attempt to the blood lipids of people by feeding a sea food diet (rich in ω3 fatty acids), some results of which are presented in table 10.5. These data support the possibility of influencing the amount and quality of lipid through diet. This is much easier to do when the person has a metabolic tendency to change. The amount and type of lipid found in the blood is partly under genetic control. This means that dietary changes are doomed to have only limited effects on the blood lipids of the population as a whole.

Table 10.5 The effects of a diet rich in sea food on lipid components in the blood of normal women and patients with insulin dependent diabetes [173, 202].

Determination	Control values (mg/dl)	Three months of treatment	
		Before (mg/dl)	After (mg/dl)
VLDL + LDL			
Cholesterol	109.0±18.7	95.7±8.8	97.3±6.3
Triglyceride	57.6±2.0	47.5±9.0	57.0±3.5
HDL_2			
Cholesterol	20.9±9.2	11.0±5.5*	18.7±7.9
Triglyceride	4.6±1.9	3.5±0.8	2.9±1.0*
HDL_3			
Cholesterol	41.2±6.4	27.8±3.9	28.8±2.7
Triglyceride	22.3±3.1	16.3±2.5	14.4±1.9*

*Significantly different from control values, $p<0.05$

10.5 FREE RADICALS, ALTERATIVE MEDICINE, AND DIET

We depend on our diet for a continuous supply of antioxidant vitamins. Some drugs may also influence the antioxidant pool in the body. Since the antioxidant action of drugs has not been widely studied, it is not surprising that reports of natural antioxidant compounds (arginine, glucocorticoids, 2-hydroxy estradiol), drugs (antiallergics such as oxatomide or astermisol), or neuroleptics (penfluridol or pimazide) are being reported.

Plant extracts (herbals) are also reported to have antioxidant properties due to their content of polyphenols, flavonoids, and carotenes. If antioxidants can influence some diseases, it should be no surprise that herbals have been used with some success for thousands of years. Unfortunately, advertising and popular attention has gotten hold of herbals and their supposed curative properties. Because of this, unsubstantiated claims are made for their benefit. The truth is probably somewhere between the inflated claims and total dismissal that comes from other quarters.

Reactive oxygen species have been demonstrated in inflammatory conditions, such as rheumatoid arthritis. Therefore, the pain associated with rheumatoid conditions may be relieved with compounds that have antioxidant properties. Traditional medicine treats these conditions with a diet high in vegetables, fruits, and fish, all of which are high in antioxidant content.

Teas from various plants are the most widespread traditional medicine. As is the case for the use of vitamin pills to treat a condition, the use of teas require a lengthy time to have an effect. In reality, such treatments are not cures but are maintaining the body's natural defenses. Caution in the use of herbals is required. The treatments may be "natural" but some plant materials also contain alkaloids and other toxins. Knowledge is required to know the safe and possibly beneficial plants from the harmful ones. Another factor is that if an herbal contains an active ingredient, it may only be present at a certain stage of the plants development, or only in a certain part of the plant (root vs leaf, etc.), or only can be extracted by specific methods (organic extractions vs boiling in water, etc.).

10.5.1 Ginseng

Danas ginseng is one of the plants that have fascinated mankind since very early times, probably because of the peculiar shape of the roots. Its alleged curative properties were reported in China 2,500 years ago. Its uses include use as a tonic, improving mental ability, a tranquilizer, antitumoral, antiviral, and antiinflammatory actions.

Ginseng's active ingredients are found in the root. The chemical composition of the root varies with geographic area and includes 30 triterpenoid glycosidases, sapagenins, ginsen oxides, phytosterols, vitamins B, C, and E, choline, peptides, amino acids, and metal ions (iron, copper, manganese, zinc, vanadium, germanium, chromium). Most of these compounds exhibit antioxidant properties. A few reports suggest that treatment with ginseng increases HDL lipoprotein lipase and cAMP. The biosynthesis of phospholipids, corticosteroids, cholesterol, and uric acid are also increased [152, 208].

10.5.2 Garlic and Onion

There is no such thing as folk-based cooking without garlic and onion. The cooking and therapeutic properties of these plants from the Allium family are widely recognized and mentioned in old books on medicine, such as the Codex Evers (13th century). Garlic was know to have antiseptic, vasodilating, anticoagulant, hypoglycemic, and antiviral properties as long ago as Biblical and Egyptian times.

The antioxidant properties of garlic are based on several components but especially on the sulfur derivatives such as ethanthiol (C_2H_5SH), dialyl disulphide ($CH_2=CH-CH_2-S-S-CH_2-CH=CH_2$), and alycin or alyl-2-propenethiosulfinate ($CH_2=CH-CH_2-S(O)-S-CH_2=CH_2$). Alylic disulfides are also responsible for the characteristic smell and lacrimatory effect. They are also responsible for the cholesterol lowering, antilipoemic, and antithyroid properties. Oils from onion and garlic inhibit *in vitro* platelet aggregation by directly acting on prostaglandin and thromboxane synthesis. The oils have a powerful action in concentrations between 5 and 10 μg/ml, a concentration that is similar to the effective concentration of antiinflammatory and antithrombotic drugs. There effect is due to the inhibition of both enzymatic pathways of the arachidonic cascade (cyclooxygenase and lipoxygenase). The production of drugs based on garlic and onion extracts is being done by companies such as Sandoz (Switzerland) and Whithorpe (USA).

10.5.3 Germinated Grains

The great nutritive capacity of germinated cereals (wheat, barley, oats) has attracted healers for many centuries. The most widely known use of germinated grains is for wheat, which is supposed to provide general vigor.

Germinated grains offer some advantages in the form of starch, proteins, and cellulose in an assimilable form; the decrease of natural enzymatic inhibitors in the grain; and the quantitative increase in the amount of available vitamins (B_1, B_2, E, A), amino acids, phospholipid, phytosterols, and minerals (K, Mg, Ca, Mn, Zn). In highly industrialized countries, especially in the US, a large array of manufactured germinated cereals are sold.

10.5.4 Apitheripy and Propolis

The therapeutic use of propolis and honey or other bee products (pollen) is centuries old knowledge, but interest in its use persists. The International Organization of Apimondia has tried to scientifically study bee products and to provide these as medications. Pollen and its derivatives are rich in vitamins E, C, and B and in amino acids. Bee venom has phospholipase and antiinflammatory phenols. But the most interesting product of bees is propolis. This complex mixture contains phenols, waxes, anthocyans, and flavonoids. Therefore, propolis has been used in traditional medicine as an antiseptic, bactericide, and antiviral agent. Its antioxidant properties were tested by its capacity to inhibit myloperoxidase derived from leukocytes. Propolis extract exhibits antioxidant properties similar to vitamins C and E, as it quenches ESR signals of some free radicals and inhibits peroxidation of liver microsomes. Propolis is used in traditional medicine for small burns, wounds, and orally. It has no toxic effects at doses up to 1g/kg body weight.

10.5.5 Copper

The wearing of copper rings and bracelets for relief from rheumatoid disease has been practiced for a long time. The first scientific explanation of a potential benefit pointed to high frequency electromagnetic oscillations that appear at the ends of the bracelets. It is known that an electromagnetic field exerts a temporary antiinflammatory action. However, such an explanation seems unlikely, so attention was turned to biochemical explanations.

Copper in the plasma increases when a person has an infection or inflammation. The increased copper is carried in ceruloplasmin, which is able to carry up to 8 copper atoms per molecule. Ceruloplasmin is an important antioxidant and is a stress protein. The increase in copper and ceruloplasmin seems to be mediated by interleukin-1. This increase is necessary to the defense of the organism as copper is the active element in many enzymes, such as superoxide dismutase, cytochrome oxidase, dopamine-β-hydroxylase, tyrosinase, and amino acid oxidases, which are involved in the biosynthesis of peptidic hormones in hypophysis. As Sorensen has shown, the increased copper is due to the free element (usually less than 1 mg/dl) that is reactively binding to free sulphydryl groups of proteins or hemoglobin, but also with amino acids and drugs [182]. Copper salts of amino acids possess analgesic effects as they are able to penetrate the blood-brain barrier and decrease the sensation of pain. The analgesic effect of copper salts is due to the activation of receptors for opiates. Some organic complexes of copper exhibit antidiabetic, anticonvulsive, antiulcerous, and antitumor effects, but also have serious side effects.

None of this really solves the problem on copper bracelets and their supposed antiinflammotory effect. While copper has the properties described above, most scientists believe metals cannot penetrate unbroken skin. The question of if and how a copper bracelet increases copper in the blood is unknown.

10.5.6 Vegetarianism

Vegetarianism has millions of adherents in all cultures around the world. Most of these people claim mystical or religious benefits. The scientific conclusions are difficult to clearly state even though many studies have been done. This is because many different experimental conditions have been used to study this. A true, completely vegetarian diet does not use any animal product. This risks protein deficiency based on essential amino acids, and essential fatty acid deficiency. Therefore, such a strict vegetarian diet should theoretically lead to a metabolic imbalance, weakness, fatigue, and would prevent growth of children. Consequently, some protein and vitamin sources are usually accepted and most vegetarian diets allow the eating of eggs and dairy products and, sometimes, fish.

Scientist from Holland have made an interesting study about the differences between metabolic parameters of vegetarians (ovo-lacto) and those eating a meat based diet. The results of these studies showed that a balanced vegetarian diet allows the assimilation of a moderate amount of carbohydrate and an optimal amount of saturated and unsaturated fatty acids [13, 25, 63]. As shown in table 10.6, the most significant differences are in carbohydrate, fatty acid, and fiber intake, but most of these differences are not statistically valid.

10.5.7 Condiments and Spices

A recent study has confirmed the involvement of factors related to the environment in carcinogenesis. This study also confirmed the protective role of vitamins C and E as well as carotenoids. In this study, some carcinogenic factors that were examined include smoking, smoked or salted meat, and nitrosamines (for gastric cancer).

On the other hand, there are many geographic areas, such as India, South America, Bulgaria, or Northern Africa, in which food preparation involves the extensive use of spices. While it may be intuitively true that spices would be irritating to the gut and increase the incidence of ulcer or gastric cancer, the incidence of these is not greatly increased where the diet contains large amounts of spices. The reason for this is partly in the antioxidant properties of spices. Black pepper and related spices contain phenolic amides that have antioxidant properties similar to vitamin E. Therefore, if the spices on the one hand might lead to inflammation, the presence of the antioxidants prevents this from becoming a serious development.

Table 10.6 Metabolic differences between individuals with normal and vegetarian diets.

Parameter	Normal diet	Vegetarian diet	Significance ($p<$)
Energy (MJ)	11.0±1.7	11.9±1.5	NS*
Protein (g/day)	93±22	87±8	NS
Fats (g/day)	112±25	107±18	NS
Carbohydrate (g/day)	352±44	292±43	0.025
Cholesterol (mg.day)	279±12	263±16	NS
PUFA† (g/day)	16.6±4.5	25.8±9.2	0.05
Unsat. FA:Sat. FA ratio	0.18±0.09	0.30±0.12	0.05
Fiber (g/day)	33.1±9.3	48.2±5.7	0.005
Bile acids (mmol/kg/day)	12.5±4.8	8.3±2.6	NS
Serum cholesterol (mM)	4.84±0.71	4.17±0.47	NS
Serum triglyceride (mM)	1.20±0.28	1.06±0.38	NS
Vitamin C (mg/dl)	106±78	183±82	0.01
Vitamin A (mg RE‡)	1335±753	1972±1242	NS
β-Carotene (mg)	4390±3343	11940±7151	0.01
Vitamin E (mg)	11±5	25±10	NS
Zinc (mg)	10±2	11±4	NS
Selenium (mg)	73±18	27±13	

*Not significant
†Polyunsaturated fatty acids
‡ Retinol equivalents

10.5.8 Exotic Plants and Antioxidants

In the USA and Western Europe, another form of alternative medicine includes the use of exotic plants from South America, the Pacific, and Southeast Asia. Most of these plants are used as extracts, pills, or teas, and all are claimed to have beneficial biological effects, particularly to be antioxidants. The cost of these herbals is great, so it may be useful to examine their alleged antioxidant properties. As seen in table 10.7, most of these exotic plants contain a large amount of vitamin C. On the other hand, so does grapefruit, oranges, and lemons.

Similar remarks can be made about exotic dishes such as humus from the Middle East, which is rich in sesame butter, olive oil, and garlic, or Japanese sushi, which is based on sea foods that are high in calcium, potassium, iron, magnesium, vitamin C, carotene, and fiber.

Table 10.7 The antioxidant and biological properties of some exotic plants.

Plant	Antioxidant and nutrative components
Avacado (& guacamole)	Viamin A*, potassium†, ω3 fatty acid*, fiber †
Banana	Potassium†, vitamin C*, magnesium†
Bread fruit	Vitamin C†
Brazil nuts	Selenium
Carambola (star fruit)	Vitamin C†
Cherimoya (custard apple)	Vitamin C†, terpenes, flavonoids
Evening primrose oil	Linoleic acid†
Figs	Calcium†, potassium*, carbohydrate†
Guava	Vitamin C†, pectin
Kumquat	Vitamin C†
Lychee	Vitamin C†
Passion fruit	Vitamin C†, flavonoids*, vitamin A*, iron*
Persimmon	Vitamin C†, carotenes*, potassium†
Prickly pear	Calcium†, magnesium†
Sugar apple (sweetsop)	Vitamin C†
Unique fruit (hybrid)	Vitamin C†

*Moderate
†High

10.6 ANTIOXIDANT THERAPY

As presented throughout this book, antioxidants within the body provide a defense system. Their main role is the defence of the physiologic activity of the cells by protecting against the harmful formation or spread of reactive oxygen species. It has been pointed out that many diseases involve oxidative stress that aggravates the course of the disease. Given this is true, it should be apparent that antioxidant supplementation cannot cure disease, but it may help limit its extent and aid in the cure provided by other drugs and the body's natural healing ability.

The question remains of when to administer antioxidants and how much to use. These are difficult questions for which there are currently no objective answers. International studies involving thousands of people have pointed out the importance of natural antioxidant nutrients (vitamins C and E and carotenes) in maintaining health and contributing to protection from pathology [49, 65, 149, 205]. A World Health Organization survey of 16 European countries showed an inverse correlation between plasma α-tocopherol levels and mortality from ischemic heart disease. A prospective study established an association of deficient levels of carotene and vitamin C with a high mortality from ischemic heart disease and stroke. The Harvard Health Professionals study [13] investigated 40,000 healthy men and 80,000 healthy women and found that a supplement of at least 100 IU of vitamin E per day for a minimum of two years resulted in a 40% reduction of risk for heart disease. This study did not prove a causal relationship, but provide evidence of an association between a high intake of vitamin E and a lowering of coronary vascular disease.

A team studying free radicals at Guy's Hospital in London showed that a supplementation of vitamin E (300 mg), vitamin C (250 mg), and β-carotene (15 mg) increased the total antioxidant content in plasma as well as the tocopherol:LDL cholesterol ratio. Few think that antioxidant pills should replace fruits and vegetables, but for people undergoing oxidative stress, supplementation might be a viable treatment.

Chronic inflammatory disease involves a large number of activated phagocytes, the greatest source of reactive oxygen species. For example, joint disease with chronic inflammation is accompanied by the formation of approximately 500 nmol. Hydrogenperoxide/minute in the inflamed joint [29, 193]. In such a case, supplementation with antioxidants combined with plenty of fruits and vegetables may support the body's defenses and limit tissue damage.

The use of natural or synthetic antioxidants needs to be appropriate to the condition against which protection is sought. It must be remembered that each antioxidant has a safe concentration range and a tolerated one (not having serious side effects). The safe and tolerated ranges are strongly dependent on structure. For example, vitamin C and carotenes might be used without toxic effects (except for gastric irritation), while vitamin A, selenium, and NSAIDs have clear-cut toxic limits and should be administered with caution.

The question of when to administer antioxidants is also a difficult question and largely depends on the individual. Generally, in the first days of the acute phase of a disease the administration of antioxidants does not help. Following the critical period, administration of antioxidants may limit the extent of reactive oxygen species induced damage and help the immune and biologic defenses to act efficiently.

CITED REFERENCES

1. Aboobaker AD, Balgi D. (1995) Flavonoids; antioxidative protection against aflatoxins in vivo. *FASEB J.* 8: 1095-1099.
2. Afana Ev IB. Superoxide ion: Chemistry and biological implications. *CRC Press*, Boca Raton, FL.
3. Albanes D, Heinonen OP, et al. (1995) Effects of tocopherol and carotenes on cancer incidence. *Am J Clin Nutrition* 64: 1427S-1430S.
4. Ames BN. (1985) Dietary carcinogens and anticarcinogens. *Science* 227: 375-381.
5. Ames BN (1989) Endogenous oxidative stress in DNA damage ageing and cancer. *Free Radical Res Com* 7: 121-128.
6. Arlias IN, Jacoby WB. (1987) *Gentathione: metabolism and function*. Raven Press, NY.
7. Aruoma OI, Halliwell B, Hoey B. (1989) The antioxidative action of N-acetyl-cysteine. *Free Radical Res Com* 6: 593-598.
8. Aruoma OI, Halliwell B. (1991) *Free radicals and food additives*. Taylor and Francis, London.
9. Aruoma OI. (1992) *Experimental research tools in free radicals in tropical diseases. Ed*: Aruoma, OI. Harwood Acad. Publ., London. Pg 233-267.
10. Aust AE Lund LG. (1993) *The role of iron in asbestos-catalyzed damage to lipids and DNA*. Ed: Reddy CC, Hamilton GA, Madastha KN. Acad Press, San Diego, CA. Pg. 567-605.
11. Aviram M. (1995) Oxidative modifications of LDL and its relation to atherosclerosis. *Israel J Med Sci* 31: 241-249.
12. Awashti JC, Zimniak P, Singhal SS. (1995) Physical role of GSH transferase in protection against lipid peroxidation. *Biochem Arch* 11: 47-54.
13. Baker H, De Angeli SB, Franck O. (1996) Human plasma patterns during 14 day investigation of vitamin E, C and carotenes and various combinations. *J AM Coll Nutr* 15: 159-163.
14. Barja G (1993) Oxygen radicals; a failure or a success of evolution. *Free Rad Res Com* 63.
15. Bauchard M, Vian C. (1995) Benzopyrene diolepoxides of hemoglobin adducts. *Arch Toxicol* 69: 540-546.

16. Baynes JW, Monnier VM (1990) The maillard reaction in diabetes and nutrition. *Free Rad Res Com* 9: 65-67.
17. Berg J, vanden Kamp JA, Lubin BH (1992) Kinetics and specificity of hydroperoxide-induced oxidative damage in red blood cells. *Free Rad Biol Med* 12: 487-498.
18. Bhlock G (1991) Vitamin C and cancer prevention, the epidemologic evidence. *Am J Clin Nutr* 53: 270S-282S.
19. Bigalow JE, Varnes ME (1989) Role of GSH and other thiols in cellular response to radiation and drugs. *Drugs Metab Rev* 20: 1-12.
20. Bianann AD, Maxwell RJ, Miller JP (1995) Antioxidants, von Willebrand factor, and endothelial cell injury in hypercholesterolemic and vacular disease. *Atherosclerosis* 116: 191-198.
21. Bindoli A, Rigobello MP, Deeble DJ (1992) Biological and toxicological properties of the oxidation products of catecholamines. *Free Rad Biol Med* 13: 391-406.
22. Black G, Menkes M (1983) *Nutrition and cancer prevention*. Ed: Moon TE, Micazzi MS. M. Dekker, Inc., NY. Pg. 341.
23. Blomhoff R (ed.) (1995) *Vitamin A in health and disease*. M. Dekker, Inc., NY.
24. Bonanome A, Biasia F, Pradela M, Pagnan A (1996) n-3 fatty acids do not enhance LDL susceptibility to oxidation in hypertriacylglycerolemia in hemodialyzed subjects. *Am J Clin Nutr* 63: 261-266.
25. Brady WE, Perlman JA, Bowen P (1994) Human serum carotenoid concentrations are related to physiologic and life style. *Immunol Today* 15: 7-10.
26. Brenan LA, Hannigen BM, Barnett YA (1996) The effect of antioxidant supplementation on theoxidative-induced stress response in lymphocytes. *Biochem Soc Trans* 24: 755-758.
27. Butler J, Honey BM (1992) Reactions of glutathione and thyil radicals with benzoquinones. *Free Rad Biol Med* 12: 337-346.
28. Buttke TM, Sandstrom PA (1994) Oxidative stress as a mediator of apoptosis. *Immunol Today* 15: 7-10.
29. Cadenas E (1988) Biochemistry of oxygen toxicity. *Ann Rev Biochem* 58: 79-110.
30. Cadenas E, Packer L (1995) *Handbook of antioxidants*. M. Dekker, Inc., NY.
31. Cavalieri EL, Rogan EG (1995) Central role of radical cations in metabolic activation of polyaromatic hydrocarbons. *Xenobiotica* 25: 677-688.
32. Ceballos I, Agid F, Nicole A (1990) Neurodegenerative disorders due to brain antioxidant system deficiencies. In: *Advances in experimental biology and medicine*, Ed: Emerit I, Packer L, Auclair C. Plenum Press, NY, Pg 193-198.
33. Cheng PW, Boat TF (1995) Differential effects of ozone on lung epithelial fluid volume and protein content. *Exp Lung Res* 21: 351-365.
34. Cimino M (1989) Generation of H_2O_2 by brain mitochondria. *Arch Biochem Biophys* 269: 623-638.
35. Connor HD, Lacagnin LB, Knecht KT (1990) Reaction of glutathione with a free metabolit of carbon tetrachloride. *Mol Pharmacol* 37: 443-451.

36. Cores GA (1995) Detection of oxidative stress in heart by estimating malondialdehyde by HPLC. *J Molec Cariol* 27:1645-1649.
37. Correa P (1985) *Diet and human cancerogenesis*. Ed: Hill JV, Gebors MJ. Elswier, NY, Pg 109.
38. Creutzenberg O, Bellmann B, Klingebil B (1995) Ozone increases phagocytosis and chemotaxis. *Exp Toxicol Pathol* 47: 202-209.
39. Cross CE (1987) Oxidative stress and antioxidants. *Annuals Int Med* 107: 530-542.
40. Cunningham J, Leffell M, Mearkle P (1995) Elevated plasma ceruloplasmin in insulin-dependent diabetes: evidence for increased oxidative stress as a variable complication. *Metabolism* 44: 996-999.
41. Cutler RG (1985) Antioxidants and longevity of mammalian species. In: *Molecular biology of aging*. Ed: Woodhead AD, Blackett PD, Hollender A. Plemum Press, NY, Pg 15-73.
42. Cutler RG (1989) Antioxidants and life span. In: *Physiology of oxygen radicals*. Ed: Taylor AE, Ward P. APS Publ, Bethesda.
43. Cutler RG (1991) Antioxidants and aging. *Am J Clin Nutr* 53: 373S.
44. Davies MB, Partridge DA (1993) *Vitamin C; its chemistry and biochemistry*. Royal Soc Chem. Cambridge.
45. Dawson RJ, Beal MF, Bondy SC (1995) Excitotoxins, aging and environmental neurotoxins. *Toxicol Appl Pharmacol* 134: 1-17.
46. Decker EA (1995) The role of phenolics, conjugated linoleic acid, carnosine and pyrroloquinonine quinones as nonessential dietary antioxidants. *Nutr Rev* 53: 49-58.
47. Delrio LA, Sandalio LM (1992) Metabolizm of reactive oxygen species in peroxisomes and cellular implications. *Free Rad Biol Med* 13:557-580.
48. Dimonte DS (1984) Alterations of intracellular thiol homeostasis during metabolism of menadione in liver. *Arch Biochem Biophys* 235: 334-342.
49. Diplock AT, Machlin LJ, Packer L (Ed) (1984) *Vitamin E biochemistry and health implications*. New York Acad Sci, vol 570.
50. Diplock AT (1995) Safety of antioxidant vitamins and carotenes. *Am J Clin Nutr* 62: 1510S-1516S.
51. Dizardoglu M (1991) Chemical determinations of free radical-induced damage to DNA. *Free Rad Biol Med* 10:225-242.
52. Doelman CJ (1989) *Oxidative stress and antioxidants in antioxidant therapy and preventive medicine*. Ed: Everit I, Auclair C. Plenum Press, NY, Pg. 3-16.
53. Dolal NS, Suyan MN, Wilhelm R (1989) Oxygen free radicals are invovled on coal-induced silicosis. *Am Occup Hygiene* 33: 79-89.
54. Downey M, Yellon DM (1992) Implication of free radicals in reperfusion injury. In: *Myocardial protection in phathophysiology of reperfusion injury*. Ed: Yellon DM, Jennings RB. Raven Press, NY, chapter 3.
55. Duthie GG, Robertson DD, Maughan RY (1990) Blood antioxidant status and erythrocyte lipid peroxidation following distance running. *Arch Biochem Biophys* 282: 78-83.

56. Ernst JS, Lichtenstein AH, Contois JH (1996) Effects of National Cholesterol Education Program step 2 diets related to fish derived fatty acids and plasma lipoproteins. *Am J Clin Nutr* 63: 234-241.
57. Elinder LS, Hadell K, Johansson JJ (1995) Probucal treatement decreases concentration of diet-derived antixodants. *Athero Thromb Vasc Biol* 15: 1057-1063.
58. Emerit I (1994) Reactive oxygen species, chromosome mutations and cancer; possible role of clastogenic factors in carcinogenesis. *Free Rad Biol Med* 16: 99-109.
59. Esterbaur H, Gebicki J, Puhz H, Jurgens SG (1992) The role of lipid peroxidations of LDL phospholipids. *Free Rad Biol Med* 73: 341-348.
60. Forman HJ, Liu RM (1995) Glutathione synthesis in oxidative stress. *Antioxidants, Health Disease* 2: 189-212.
61. Forrest VJ, Kang YH McClain DE (1994) Oxidative stress induced apoptosis prevented by Trolox. *Free Rad Biol Med* 16:675-684.
62. Frei B, Ames BN (1988) Antioxidant defense and lipid peroxidation in blood plasma. *Proc Nat Acad Sci* US 85: 9748-9753.
63. Freddy K, et al. (1990) Antioxidant levels in blood plasma of cardiovascular patients. *Biochem Soc Trans* 18:1041-1047.
64. Galli C, Simopoulos AP (1991) Dietary n-3 and n-6 fatty acids. *Life Science*, Plenum Press, NY.
65. Gerlach M, Riederer P, Youdim MB (1995) Neuroprotective therapeutic strategies. *Biochem Pharmacol* 50: 1-16.
66. Gimpel JA, Lahpor JP, Molen AJ (1995) Reduction of reperfusion injury of human myocardium by allopurinol; a clinical study. *Free Rad Biol Med* 19: 251-255.
67. Green LS (1995) Asthma and oxidative stress, nutritional environmental and genetic risk factors. *J Am Col Nutr* 14: 317-324.
68. Gutierez MC, Bucio L, Sousa V (1995) The effect of chronic and acute ethanol administration on morphology and lipid peroxidation. *Human Exper Toxicol* 14: 324-334.
69. Gutteridge JMC (1993) Free radicals in disease processes; a compilation of cause and complication. *Free Rad Res Com* 19: 141-158.
70. Halliwell B, Hault JRS, Blake DR (1988) Oxidants, inflammation and antiinflammatory drugs. *FASEB J* 2: 2867-2873.
71. Halliwell B (1988) Oxygen radicals and tissue injury. *FASEB*, Bethesda.
72. Halliwell B, Gutteridge JMC (1998) *Free radicals in biology and medicine*, 2nd edition. Clarendon Press, Oxford.
73. Halliwell B (1990) How to characterize a biological antioxidant. *Free Rad Res Com 9:* 1-13.
74. Halliwell B (1991) Drug, antioxidant effects; a basis for drug selection? *Drugs* 42: 569-605.
75. Halliwell B (1992) Oxygen radicals as key mediators. *Chem* 57:2140-2143.

76. Halliwell B, Cross CE, Guttaridge JMC (1992) Free radicals, antioxidants and human diseases; where are we now? *Lab Clin Med* 119: 598-620.
77. Harman D (1994) Aging; prospects for increses in the functional life span. *Age* 17: 119-146.
78. Hassan HM (1984) Formation of free radicals in pathological conditions. In: *Free radicals in molecular biology, aging, and disease.* Ed: Armstrong D, Slater TF. Raven Press, NY. Pg 77-83.
79. Hastings TG, Lewis DA, Zigmond MJ (1966) Role of oxidation in the neurotic effects of intrastitial dopamine injections. *Proc Nat Acad Sci* US 93: 1956-1961.
80. Hayashi O, Niki E, Kondo M, Yoshikawa T (1988) *Medical, biochemical and chemical aspects of free radicals*, vol 2. Elsevier Publ., Amsterdam.
81. Heinonen OP, Albanes D (1994) The effect of vitamin E and b-carotene on the incidence of lung cancer in male smokers. *New Engl J Med* 330: 1029-1035.
82. Henschler R, Glatt HR (1995) Induction of cytochrome P450-A in hematopoetic stem cells by hydroxylated metabolites of benzend. *Toxicol In Vitro* 9: 453-458.
83. Hong YL, Pan HZ, Scott MD (1992) Activated oxygen generation by a primaquine metabolite; inhibition by antioxidants from Chinese herbal remedies. *Free Rad Biol Med* 12: 213-218.
84. Hochstein P, Rice-Evans C (1982) Lipid peroxidation and membrane alterations in erythrocyte survival in lipid peroxides. In: *Lipid peroxides in biology and medicine.* Ed: Yagi K. Academic Press, Orlando. Pg 81-88.
85. Husen NS (1994) New biological and clinical roles for the n-6 and n-3 fatty acids. *Nutr Rev* 52: 162-167.
86. Hyslop PA, Hinshaw DB, Vosbeck D (1995) Hydrogen peroxide as a patent bacteriostatic antibiotic; inplications for host defence. *Free Rad Biol Med* 19: 31-37.
87. Janssen YM, Bachavsky A, Treadwell M (1995) Crocidolete binding to proteins and DNA. *Proc Nat Acad Sci* US 92: 8458-8462.
88. Kaufmann WK, Paules RS (1996) DNA damage and cell cycle check-points. *FASEB J* 10: 238-247.
89. Klebanoff SJ (1988) *Phagocytic cell products of oxygen metabolism. In: Inflammation; basic principles and chemical correlates.* Ed: Gallin JI, Goldstein IM, Snyderman R. Raven Press, NY. Pg 391-444.
90. Koob M, Dekant W (1991) Bioactivation of xenobiotics by formation of toxic glutathione cosyngates. *Chem Biol Interact* 77: 107-136.
91. Koren HS, Bromberg PA (1995) Respiratory distrubances in asthmatics due to ozone. *Int Arch Allergy Immunol* 107: 236-239.
92. Kirinski NI (1984) *Oxygen free radicals are formed in photosensitized reactions. In: Oxygen radicals and Biology.* Ed: Bors W, Saran N. De Gruyter Publ, Berlin. Pg 453-468.
93. Kulkarn AP, Murty KR (1995) Xenobiotic metabolism in human peroxidase-mediated oxidation and bioactivation of 2-aminofluorene. *Xenobiotica* 25: 799-810.

94. Kummerow FA (1995) On a nutritionally sound-modified fat diet. *Pediatric Asthma, Allergy, Immunol* 10: 5-8.
95. Kuzuya M, Kuzuya F (1993) Probucol as an antioxidant, antiatherogenic drug. *Free Rad Biol Med* 14: 67-78.
96. Lagercrantz C (1992) Penicillin is inactivated by hydrogen peroxide. *Free Rad Biol Med* 13: 455-458.
97. Lapenna D, Gioia SD, Mezzetti A (1995) The prooxidant properties of captopril. *Biochem Pharmacol* 50: 27-32.
98. Leng GC, Horrobin DF, Fowkes FG, Ellis DK (1994) Plasma essential fatty acids; cigarette smoking and dietary antioxidants in peripheral arterial disease. *Atherosc Thromb* 14: 471-478.
99. Lelieveld J (1990) Influence of cloud's photochemical processes on troposphereic ozone. *Nature* 343: 6255-6259.
100. Leng GC, Horrobin DF, Fowkes FG, Ellis DK (1994) Plasma essential fatty acids; cigarette smoking and dietary antioxidants in peripheral arterial disease. *Atherosc Thromb* 14: 471-478.
101. Levin M, Dhariwal KR, Welch RW (1995) Determination of optimal vitamin C requirements in humans. *Am J Clin Nutr* 1347S-1355S.
102. Lin Y, Jamieson D (1995) Role of neutrophiles in hyperbaric and normobaric oxygen toxicity in cigarette smoking. *Pathophysiology* 2: 9-16.
103. Lissi EA, Videla LA, Boveris A (1991) *Metabolic regulation in oxidative stress. In: oxidative damage and repair; chemical, biological, and medical aspects*. Ed: David D. Pergamon Press, Oxford. Pg 444-448.
104. Liu L, Wells PG (1995) Teratogens like phenytoin are activated by peroxidase. *Toxicol Appl Pharmacol* 134: 71-80.
105. Luft R, Landau BR (1995) Mitochondrial medicine. *J Int Med* 328: 405-442.
106. Machlin LJ (1988) *Influences of antioxidant vitamins on cataract formation. In: Medical, biochemical, and chemical aspects of free radicals*. Ed: Hayashi O, Niki E, Kondo M, Yoshikawa Y. Elsevier, Amsterdam. Pg 351-360.
107. Madhavi DL, Deshpande SS, Salumkhe DK (ed) (1995) *Food antioxidants; technology, toxicology, and health perspectives*. M. Dekker, Inc., NY.
108. Mannaioni PF, Bello MG, Rasputin S (1995) Activation of xenobiotics into free radicals by prostaglandin synthase and liver microsomes. *Adv Prostagl Thrombox Leuko Res* 23: 215-218.
109. Marcus AJ, Hajjar DP (1993) Vascular transcellular signaling. *J Lipid Res* 34: 2017-2030.
110. McLarty JL, Holiday DB, Giarrdi WM (1995) b-carotene, vitamin A and lung cancer; chemopreventive results of an intermediate endpoint study. *Am J Clin Nutr 62:* 1431S-1438S.
111. Meerson FZ (1984) *Adaptation and stress, prophylaxis*. Springer Publ., Berlin.
112. Messina MJ (1991) Oxidative stress and cancer: methodology application for human studies. *Free Rad Biol Med* 10: 177-184.
113. Neyskens FL, Prasad KD (ed) (1980) *Vitamins and cancer; human cancer prevention by vitamins and micronutrients*. Humana Press, Clifton, NJ.

114. Miquel J, Quintanilha AT, Weber H (1989) *Handbook of free radicals and antioxidants in biomedicine (3 volumes).* CRC Press, Boca Raton, FL.
115. Mistry N, Herbert KE, Evans MD (1995) Immunological detection of reactive oxygen species damage to DNA. *Biochem Soc Trans* 23: 482-486.
116. Mueller MF, Madropoulos V, Bolt HM (1995) Toxicological aspects of estrogen-mimic xenobiotics in the environment. *Eco Toxicol News* 2: 68-73.
117. Murata N, Hiroyuki H, Miromata N (1995) Formation of mutagen, glyoxal, from DNA treated with wxygen free radicals. *Carcinogenesis* 16: 2251-2255.
118. Naarala JT, Loikkanen JJ, Savolainen KM (1995) Lead amplifies glutamate-induced oxidative stress. *Free Rad Biol Med* 19: 689-693.
119. Nensetter MS, Halvorsen B, Resvold O (1995) Paracetamol inhibits copper and mononuclear cell-mediated oxidation of low density lipoproteins. *Arterioscl Thromb Vasc Biol* 15: 1338-1344.
120. Neve J, Favier A (1989) *Selenium in biology and medicine.* Degruyer Publ., London.
121. Oberley LW (1982) *Superoxide dismutase.* CRC Press, Boca Raton, FL.
122. Oberley TD, Schultz JL, Oberley LW (1995) Antioxidant enzyme level as a function of growth state in cell culture. *Free Rad Biol Med* 19: 53-65.
123. Olinescu R (1982) *Peroxidation in chemistry, biology and medicine* (in Romanian). Ed stiintifica, Bucuresti.
124. Olinescu R, Radaceanu V, Nita S (1992) Variations of peroxides and total antioxidants in plasma of normal donors as a function of sex, age and blood group. *Rev Roum Med Int* 30: 201-206.
125. Olinescu R, Constantinescu V (1992) Variations of the phagcytic capacity of leukocytes and total antioxidants in blood of patients with pneumoconisis and lung cancer. *Rev Roum Med Int* 30: 45-49.
126. Olinescu R, Alexandrescu R, Nita S (1985) Comparative studies concerning lipid peroxidation in experimental ethanol-induced liver injury and liver diseases. *Rev Roum Med Int* 22: 293-300.
127. Olinescu R, Smith T, Hertoghe J (1997) *The body's battle against pollution.* Nova Science Publishers, Commack, NY.
128. Olinescu R, Alexandrescu R, Hula S, Kummerow FA (1994) Tissue lipid peroxidation may be triggered by increased formation of bilirubin. *Res Com Chem Pathol Pharmacol* 84: 27-34.
129. Packer L (1984) Oxygen radicals in biological systems. *Methods in Enzymology* 105: 4-32.
130. Pahan K, Smith BT, Singh I (1996) Epoxide hydrolase in human and rat peroxisomes; implicatons for disorders of peroxisomal biogenesis. *J Lipid Res* 37: 159-167.
131. Parola M Leonarduzzi G, Robino G (1996) On the role of lipid peroxidation in the pathogensis of liver damage induced by long standing cholestasis. *Free Rad Biol Med* 20: 351-360.
132. Parks DA, Bulken GB, Granger DN (1983) Role of oxygen free radicals in shock, ischemia and organ preservation. *Surgery* 94: 427-432.

133. Parks DA, Granger DN (1983) Role of oxygen free radicals in disease. *Surgery* 94: 415-422.
134. Passwater RA (1983) *Selenium as food and medicine.* Oxford.
135. Poot M, Rabinovich PS, Hoenhn H (1989) Free radical-mediated cytoxicity of desferrioxamine. *Free Rad Biol Med* 6: 323-338.
136. Poot M (1991) Oxidants and antioxidants in proliferative senescence. *Mutation Res* 256: 177-189.
137. Poulsen He, Loft S (1995) Early biochemical markers of effects, enzyme induction, oncogene activation, markers of oxidative damage. *Toxicol* 101: 55-64.
138. Prasad AS (1996) Zinc deficiency in women and children. *J Am Col Nutr* 15: 113-120.
139. Prestera T, Talalay P (1995) electrophile and antioxidant regulation of enzymes that detoxify carcinogens. *Proc Nat Acad Sci* US 92: 8965-8969.
140. Pryor WA (ed) (1976-1984) *Free radicals in biology, volumes I - VI.* Academic Press, NY.
141. Pryor WA, Frey B (ed) (1994) *Free radicals and natural antioxidants in human health and disease.* Academic Press, NY.
142. Pryor WA, Godber SS (1991) Noninvasive measurement of oxidative stress in humans. *Free Rad Biol Med* 10: 177-184.
143. Quinlan GJ, Evans TW, Gutteridge JMC (1994) Oxidative damage to plasma proteins in adult respiratory distress syndrome. *Free Rad Res* 20: 288-306.
144. Quiroga GB (1992) Brown fat thermogenesis and exercise; two examples of physiolgical oxidative stress. *Free Rad Biol Med* 4: 325-341.
145. Radomski M, Salos E (1995) Nitric oxide, biological mediator, modulator and factor of injury; its role in the pathogenesis of atherosclerosis. *Atherosclerosis* 118: 569-580.
146. Rattan IS, Sibaska GE, Wikmar FP (1995) Levels of oxidative DNA damage product 8-hydroxy-2- deoxyguanosine in human serum increases with age. *Med Sci Res* 23: 468-471.
147. Reiter R (1995) *Melatonin; your body's natural wonder drug.* Bantam.
148. Reed DJ, Oxidative stress and mitochondrial perimeability transition. *Antiox Health Dis* 2: 231-263.
149. Rice-Evans C, Diplock AT (1993) Current status of antioxidant therapy. *Free Rad Biol Med* 15: 77-96.
150. Roberfroid M, Calderon PB (1995) *Free radicals and oxidative processes in biological systems.* M. Dekker, Inc., NY.
151. Roberts L, Wood DA, Lampe FC (1995) Trans-isomers of oleic and linolenic acids in adipose tissue and sudden cardiac death. *Lancet* 345: 278-281.
152. Rong Y, Geng Z, Lau BH (1995) Gingko biloba attenuates oxidative stress in macrophages and endothelial cells. *Free Rad Biol Med* 20: 121-128.
153. Rudman D (1991) *Health effects of omega-3 fatty acids in sea food. In: Biology of fatty acids.* Ed: Simopoulos AP. Krager Publ., Berlin. Pg 143-154.
154. Sakar M, Yamagani K (1995) Xanthine oxidase mediates paraquat induced toxicity to endothelial. *Pharmacol Toxicol* 77: 36-46.

155. Sakaguki S, Yokoto S (1995) Endotoxin-induced hepato-toxicity involves activation of phagocytising leukocytes. *Pharmacol Toxicol* 77: 81-86.
156. Samuni A, Cook J, Black CD (1991) On free radical production by PMA-stimulated neutrophils as monitored by luminol-amplified chemiluminescence. *Free Rad Biol Med* 10: 305-312.
157. Sanzgiri UJ, Kim HJ, Dallas CE (1995) Effects of route and pharmacokinetics of carbon tetrachloride. *Toxicol Appl Pharmacol* 134: 148-154.
158. Sarafian TA, Breden OE (194) Is apoptosis mediated by reactive oxygen species. *Free Rad Res* 21: 1-8.
159. Sarker AH, Watanabe S, Seki S (1995) Superoxide radical-induced single strand DNA breaks and repair of the damage in cell free system. *Mutat Res* 337: 85-95.
160. Sato M, Bremner I (1993) Oxygen free radicals and metallothionein. *Free Rad Biol Med* 14: 339-342.
161. Schoenberg MH, Birk D, Beger HG (1995) Oxidative stress and chronic pancreatitis. *Am J Clin Nutr* 62: 1306S-1314S.
162. Schwartz JL, Antoiades DZ, Zhao SC (1990) Molecular and biochemical programming of oncogenesis through activity of pro-oxidants and antioxidants. *Arch Biochem Biophys* 282: 262-245.
163. Seeger W, Hasen T, Roesseg R (1995) Hydrogen peroxide-induced increased release in lung endothelial and epithelial permeability. *Microvasc Res* 50: 1-17.
164. Semaccini C, Lial I (1994) LDL modification by activated leukocytes; a cellular model of mild oxidative stress. *Free Rad Biol Med* 16: 49-55.
165. Sharp IS, Benowitz NL, Beswirk AD (1995) Tobacco smoke increases platelet aggregation. *Thromb Hemostasis* 74: 730-735.
166. Shaw VH, Spencer JD (1994) Oxidative and reductive roles of iron and protective role of vitamin C. *8th Symposium Mol Biol Haematopoesis Proc.* Pg 539-554.
167. Sies H (ed) (1985) *Oxidative stress.* Academic Press, NY.
168. Sies H (1993) Stragegies of antioxidant defense. *Eur J Biochem* 215: 213-219.
169. Sies H, Krinsky N (1995) The present status of antioxidant vitamins and b-carotenes. *Am J Clin Nutr* 62: 1298S-1300S.
170. Sies H, Stahl W (1995) Vitamins E and C and carotenoids as antioxidants. *Am J Clin Nutr* 62: 1315S-1321S.
171. Siems WG, Van Kuijk F, Maas R (1994) Uric acid and glutathione levels during short term whole body exposure to stress. *Free Rad Res Com* 19: 299-303.
172. Simmoneli F (1989) Lipid peroxidation and cataractogenesis in diabetes and severe myopia. *Exp Eye Res* 49: 181-188.
173. Simopoulos AP (1991) Omega-3 fatty acids in health and disease, in growth and development. *Am J Clin Nutr* 54: 438-463.
174. Sinet PM, Ceballos-Picot I (1992) Role of free radicals in alzheimers disease and downs syndrome. In: *Free Radicals in Brain Processes.* Ed: Packer L, Prilipko L. Christen J. Springer Verlag, NY. Pg 91-98.

175. Singh SV, Rahman Q (1987) Interrelation between hemolysis and lipid peroxidation of human erythrocytes induced by silicate dust. *J Appl Toxicol* 7: 91-96.
176. Singh SV, Jobal J (1990) Cytochrome P450 reductase, antioxidant enzymes and cellular resistance to doxorubicin-resistant cells. *Biochem Pharmacol* 40: 385-387.
177. Slater TF (1982) *Free radicals, lipid peroxidation and cancer.* Academic Press, NY.
178. Slater TF (1985) *Biochemical mechanism of liver injury.* Academic Press, NY.
179. Smith TL, Kummerow FA (1989) Effect of dietary vitamin E on plasma lipids and atherogenesis in restricted ovulator chickens. *Atherosclerosis* 75: 105-109.
180. Sonal RS, Slater TF (1985) *Free radicals in molecular biology and aging.* Raven Press, NY.
181. Sohal RS, Agarwal S, Sohal BN (1995) Oxidative stress and aging in the mongolian gerbil. *Mech Age Develop* 81: 15-25.
182. Sorenson RJ (1989) Organic copper-complexes with SOD-like activity. In: *Progress in medical chemistry.* Ed: West GP, Elias CV. Elsevier, NY. Pg 152-178.
183. Spector A (1995) Oxidative stress-induced cataract; mechanism of action. *FASEB J* 9: 1173-1182.
184. Stocker R, McDonagh AF, Ames BN (1990) Antioxidant activities of bile pigments; biliverdin and bilirubin. *Methods Enzymol* 186: 301-309.
185. Szent-Gyorgyi A (1976) *Electronic biology of cancer.* M. Dekker, Inc., NY.
186. Tanaka J, Miki M, Yasuda H (1991) The effect of a-tocopherol on the oxidation of membrane protein thiol induced by free radicals generated in different sites. *Arch Biochem Biophys* 285: 344-350.
187. Tarr M, Samson F (1993) *Oxygen free radicals in tissue damage.* Birkenhauser, Boston. Chapters 6-8.
188. Toohey L, Harris MA, Allen KG (1996) Plasma ascorbic acid concentrations are related to cardiovascular risk factors in Afro-Americans. *J Nutr* 126: 121-128.
189. Toyokuni S (1996) Iron-induced carcinogenesis; the role of redox regulation. *Free Rad Biol Med* 20: 553-566.
190. Trump FB, Berezesky IK (1995) Calcium-mediated cell injury and cell death. *FASEB J* 9: 219-228.
191. Utrecht JP (1990) Drug metabolism by leukocytes; its role in drug-induced idiosyncratic reactions. *Crit Rev Toxicol* 20: 213-235.
192. Uri N (1961) Physico-chemical aspects of autooxidation. In: *Autooxidation and antioxidants, vol. 1.* Ed: Lundberg WO. Interscience, NY. Pg 55-106.
193. Varma SD (1990) Excretion of hydrogen peroxide in human urine. *Free Rad Res Com* 8: 73-78.
194. Van Der Vliet A, Bast A (1992) Role of reactive oxygen species in intestinal diseases. *Free Rad Biol Med* 12: 499-514.
195. Van Der Vliet A, O'Neill CA, Eisenach JP (1995) Oxidative damage to extracellular fluids by ozone; protection by ozone. *Arch Biochem Biophys* 32: 43-50.

196. Volk T, Ioanidis F, Hense LM (1995) Endothelial damage induced by nitric oxide; synergism with reactive oxygen species. *Biochem Biophys Res Com* 213: 196-203.
197. Vonsontag C (1988) *The chemical basis of radiation biology.* Taylor and Franc's, London.
198. Wardman P, Ross AB (1991) Radiation chemistry; literature compilations. *Free Rad Biol Med* 10: 243-247.
199. Warner HR (1994) Superoxide dismutase, aging and degenerative diseases. *Free Rad Biol Med* 17: 249-258.
200. Weitzman SA, Gordon LT (1990) Inflammation and cancer; role of phagocyte-generated oxidants in cancerogenesis. *Blood* 76: 655-663.
201. Wiese AG, Pacifici RE, Davies KA (1995) Adaptation to oxidative stress in mammalian cells. *Arch Biochem Biophys* 318: 231-240.
202. Wolff SP, Zadeh JN *Consumption of dietary lipids, hydroperoxides; a possible contributory factor for atherosclerosis.*
203. Yagi K (1983) *Peroxidation in biology and medicine.* Academic Press, NY. Pg 223-240.
204. Youn YK, Lalonde C, Demling R (1992) Oxidants and the pathophysiology of burn and smoke inhalation. *Free Rad Biol Med* 12: 409-412.
205. Yu BP, Lee DW, Marler CG (1990) Mechanism of food restriction; protection of cellular homeostasis. *Proc Soc Biol Med* 193: 13-18.
206. Yu BP (1994) Cellular defenses against damage from reactive oxygen species. *Physiol Rev* 74: 139-160.
207. Zhang L, Smith MT, Bondy B (1994) Role of quinones, active oxygen species and metals in the genotoxicity of 1,2,4-benzenetriol, a metabolite of benzene. In: *Free radicals in the environment; medicine and toxicology, vol 8*. Ed: Nohl H, Esterbauer H, Rice-Evans C. Rechelieu Press, London. Pg 521-562.
208. Zhang D, Yasuda T, Yu BP (1996) Genseng extract scavenges hydroxyl radicals and protests unsaturated fatty acids from decomposition caused by iron-mediated lipid peroxidation. *Free Rad Biol Med* 20: 145-150.
209. Zhou Q, Smith T, Kummerow FA (1993) Cytotoxicity of oxysterols on cultured smooth muscle cells from human umbilical arteries. *Proc Soc Exp Biol* Med 202: 75-80.

INDEX

A

Absorption, 8
Activation, 46, 131, 153, 178
ADP, 37, 45, 46, 64
adrenal gland, 102
adrenalin, 46, 121
adriamycin, 5, 16, 61, 91, 142
adult respiratory distress syndrome, 180
affect, 79, 80, 83, 89, 103, 123, 128, 133, 135, 143, 146
aging, 7, 31, 56, 76, 81, 85, 101, 110, 121, 123, 124, 125, 126, 129, 139, 142, 152, 154, 175, 177, 182, 183
AI, 132
Alzheimer disease, 93, 153, 154
Amadori compounds, 155
amino acids, 20, 31, 39, 42, 50, 73, 100, 113, 114, 135, 137, 147, 152, 161, 166, 167, 168
amplification, 61, 62, 79, 101
androgens, 140
anorexia, 146
ANTU, 59, 60
apoptosis, 52, 174, 176, 181
ARDS, 148
aromatic compounds, 17, 18, 19, 31, 50, 55, 57, 58, 142, 158
arthritis, 64, 92, 93, 103, 116
ascorbic acid, 104, 105, 106, 117, 136, 151, 159, 182
assimilation, 114, 168
asthma, 44, 117, 145, 146
atherosclerosis, 126, 138, 140, 141, 155, 173, 180, 183
ATP, 36, 37, 38, 46, 54, 64, 66, 151, 161
attention, 3, 87, 103, 118, 140, 165, 167

B

barbiturates, 97, 151, 160
behavior, 19, 159
bilirubin, 25, 56, 67, 68, 77, 88, 97, 98, 119, 150, 179, 182
black lung disease, 144
blocking, 16, 21, 134, 150
blood pressure, 141
blurred vision, 109
bone marrow, 83, 131
brain stem, 126

C

California, 106, 141
cancer, 9, 18, 19, 51, 65, 66, 74, 76, 77, 79, 81, 84, 97, 100, 102, 103, 105, 108, 109, 110, 111, 113, 114, 115, 117, 127, 128, 135, 141, 142, 143, 145, 148, 152, 157, 162, 163, 173, 174, 176, 177, 178, 179, 182, 183
cancer, lung, 105, 108, 128, 148, 177, 178, 179
capillary permeability, 59
carcinogenesis, 53, 113, 142, 143, 168, 176, 182
cardiac muscle, 114
cardiovascular disease, 98, 102, 109, 113, 114, 117, 125, 131, 135, 136, 138, 139, 140, 148, 154, 162, 163, 164
carotenoids, 26, 107, 108, 109, 136, 168, 181
catabolic reactions, 35
catabolism, 111
catalase, 8, 29, 37, 39, 56, 58, 59, 66, 87, 92, 93, 94, 95, 97, 126, 127, 132, 153, 155, 159
cataracts, 25, 84
catecholamines, 27, 63, 98, 116, 121, 152, 174
cell death, 52, 62, 65, 128, 182
cell lysis, 39, 57, 61, 100, 120, 121
cell suspensions, 63, 89, 109
central nervous system, 160
cerebellum, 126
cerebrospinal fluid, 64, 112
ceruloplasmin, 16, 17, 90, 98, 112, 116, 132, 135, 139, 141, 148, 151, 167, 175

chemical intoxication, 99, 100, 149, 150
chemiluminescence, 25, 28, 34, 43, 47, 48, 58, 66, 150, 155, 181
Chinese, 177
chlorophyll, 25, 108
chlorpromazine, 2, 71, 80, 85
cholesterol, 25, 44, 77, 103, 105, 117, 135, 136, 139, 140, 141, 163, 166, 169, 171
chromatography, 48, 49
complexity, 1, 2, 14, 63, 74, 129, 130, 147, 157
COMT, 152
concept, 1, 53
conditioning, 120
conscious, 151
copper, 4, 27, 28, 75, 90, 91, 92, 93, 97, 104, 111, 112, 113, 115, 116, 117, 132, 134, 135, 138, 151, 154, 161, 166, 167, 168, 179
cornea, 64
correlation, 49, 60, 68, 101, 123, 126, 141, 163, 170
correlation coefficient, 141
criterion, 29, 132
crystallin, 84, 85
cyclooxygenase, 27, 45, 46, 47, 131, 134, 156, 166
cystic fibrosis, 93, 155
cytotoxicity, 46, 65, 101

D

DDC, 161
DDT, 50
defense mechanism, 37
degradation, 15, 20, 31, 32, 33, 34, 40, 48, 67, 94, 117, 131, 135, 161
diabetes, 93, 97, 112, 136, 140, 141, 147, 154, 155, 160, 163, 165, 174, 175, 181
diet, 26, 39, 51, 53, 103, 105, 106, 107, 113, 115, 118, 125, 136, 140, 142, 143, 145, 154, 157, 158, 162, 163, 164, 165, 168, 169, 178
diethyldithiocarbamate, 161
dihydroxyphenylamine, 78
DIPS, 93
DNA, 15, 16, 18, 19, 20, 30, 47, 51, 52, 65, 73, 76, 77, 80, 81, 82, 83, 88, 93, 97, 104, 105, 111, 113, 123, 124, 125, 128, 131, 135, 143, 145, 149, 173, 175, 177, 179, 180, 181
DOPA, 78, 97
dopamine, 152, 153, 154, 168, 177
double blind, 103
Down syndrome, 93
drug use, 161
Dust, 144

dysplasia, 93, 102, 109, 142

E

EDRF, 21, 52, 137
EDTA, 20, 64, 160
eicosanoids, 44, 45, 46, 133
Electron, 3, 5, 6, 9, 27, 52, 124
embryogenesis, 52
endoperoxides, 46, 47, 55, 95, 138
endothelium, 21, 95, 98, 113, 131, 137, 138, 148, 163
endothelium derived relaxing factor, 21
enzymatic redox reactions, 5, 12
epilepsy, 64, 113
epinephrine, 132
erythrocytes, 54, 55, 56, 64, 90, 92, 94, 95, 97, 111, 120, 126, 144, 147, 148, 159, 182
ESR, 4, 9, 10, 16, 19, 20, 21, 26, 30, 49, 55, 104, 120, 124, 144, 148, 150, 167
estrogen, 18, 50, 51, 88, 111, 113, 140
eukaryotes, 123
eV, 80
extracellular fluid, 112, 182

F

FAD, 12, 13, 14, 15, 66
fats, unsaturated, 163
fatty acids, 3, 7, 32, 39, 40, 42, 47, 58, 60, 61, 73, 111, 125, 136, 140, 141, 147, 162, 163, 164, 168, 169, 176, 177, 178, 180, 181, 183
Fenton reaction, 28, 29
fiber, 168, 170
fibroblasts, 41, 51, 93, 123, 128, 143
fibrosis, 138, 144, 145, 155
flavin adenine dinucleotide, 12
flavin semiquinones, 16
flavine mononucleotide, 12
flavins, 25, 26
fMLP, 40
FMN, 12, 15
formyl-methionyl-leucyl-phenylalanine, 40
free radical formation, 4, 6, 9, 10, 16, 19, 23, 31, 48, 121, 161
Frequency, 83
fundamental, 71

G

GABA, 121, 127, 146
gamma, 81, 83
gamma rays, 26, 83

garlic, 115, 166, 170
gastric cancer, 168, 169
gene, 20, 93, 141, 143
GHS, 55, 94
ginseng, 51, 118, 166
glutamate, 121, 127, 146, 151, 152, 153
glutathione, 15, 21, 33, 37, 53, 54, 55, 56, 57, 58, 61, 63, 67, 68, 75, 85, 88, 90, 93, 94, 95, 96, 97, 99, 100, 101, 102, 104, 110, 113, 114, 124, 126, 127, 141, 143, 146, 148, 149, 150, 155, 158, 159, 161, 174, 177, 181
glutathione transferase, 96, 97, 124
GPT, 58, 59
GSH, 37, 54, 57, 58, 59, 61, 94, 97, 99, 100, 101, 173, 174

H

Haber-Weiss cycle, 3, 4
Haber-Weiss reaction, 27, 28, 160
HAD reduced form, 4, 6, 13, 15, 16, 35, 36, 37, 38, 54
hallucinogens, 14
heart disease, 21, 74, 103, 106, 108, 115, 135, 140, 149, 162, 171
heart disease, ischemic, 170
heat shock, 98
heavy, 50, 138, 151
Heinz bodies, 55, 56
heme, 9, 65, 67, 98
Heme, 98
Hemocyanin, 92
hemoglobin, 4, 9, 18, 21, 27, 54, 55, 56, 57, 64, 65, 67, 92, 98, 111, 120, 123, 135, 137, 158, 159, 168, 173
hemolysis, 54, 55, 56, 102, 127, 158, 182
hemostasis, 46
hemoxygenase, 98
hepatitis, 69, 97
Herpes simplex virus, 79
high pressure liquid chromatography, 49, 124
hippocampus, 126, 153
histamine, 41, 115, 129, 130, 134, 146, 147
homeostasis, 62, 98, 121, 175, 183
hormones, 21, 44, 50, 143, 146, 168
HPLC, 48, 49, 55, 124, 175
human immunodeficiency virus, 112
humoral, 76, 131, 133
hydrogen peroxide, 3, 4, 25, 26, 27, 28, 29, 30, 35, 37, 38, 39, 40, 41, 42, 43, 44, 48, 51, 52, 58, 66, 73, 77, 79, 83, 85, 91, 93, 94, 95, 112, 123, 128, 131, 132, 137, 143, 146, 150, 152, 154, 159, 160, 177, 178, 181, 182

hydroxyl radical, 2, 20, 25, 28, 29, 30, 31, 59, 63, 65, 66, 81, 90, 111, 112, 124, 150, 155, 183
hypercholesterolemia, 140, 141
hyperoxia, 147
hypertension, 136, 138, 140
hypertrophy, 76
hypothalamus, 110, 126
hypothesis, 110, 123, 125, 142
hypothyroidism, 97
hypoxia, 84

I

immune systems, 110
immunity, 40, 76, 104, 115, 116
immunocompetent, 103, 115
inclusion, 124
India, 169
Indomethacin, 43
Indonesia, 110
industrialized countries, 127, 135, 142, 162, 167
infants, 102
infection, 40, 43, 94, 95, 97, 98, 103, 116, 126, 128, 129, 136, 139, 148, 152, 167
infectious disease, 103
inflammation, 15, 43, 53, 63, 66, 98, 112, 114, 116, 117, 129, 130, 131, 132, 133, 134, 135, 136, 137, 138, 143, 156, 157, 167, 169, 171, 176
inflammatory, 43, 45, 46, 64, 98, 112, 116, 117, 120, 130, 131, 132, 133, 135, 139, 144, 145, 148, 152, 155, 156, 165, 171
inflammatory diseases, 120
inhibition, 40, 46, 92, 97, 104, 109, 115, 137, 166, 177
inhibitor, 44, 91, 125, 132, 161
inhibitors, 6, 8, 137, 167
insecticides, 50
insight, 150
insulin, 134, 154, 165
integrity, 43, 54, 102
interference, 161
Interleukin, 135
invasion, 155
ion, 3, 27, 28, 112, 134, 155, 173
ionization, 81
ionizing radiation, 4, 7, 8, 15, 20, 27, 29, 31, 52, 53, 80, 81, 82, 83, 84, 142, 143
ions, 4, 8, 9, 20, 24, 28, 30, 31, 33, 34, 44, 46, 58, 64, 65, 79, 81, 83, 90, 91, 104, 112, 113, 117, 132, 133, 146, 150, 151, 160, 166
iron, 4, 9, 14, 20, 27, 34, 35, 63, 64, 65, 90, 91, 104, 111, 112, 113, 115, 117, 126, 132, 138,

144, 145, 150, 151, 160, 161, 166, 170, 173, 181
Iron, 3, 63, 64, 65, 92
iron chelators, 160
iron complex, 65, 161
irradiation sickness, 20
ischemia, 14, 15, 60, 62, 63, 66, 67, 92, 136, 138, 147, 153, 160, 161, 179
isoenzymes, 96
Israel, 173
Italy, 140

J

Japan, 115, 164
Japanese, 51, 113, 124, 163, 170
jaundice, 68, 77, 79, 102, 120

K

kidney, 15, 38, 97, 99, 106, 113, 138, 155, 160
knowledge, 8, 51, 73, 75, 80, 167

L

LDL, 51, 98, 103, 105, 111, 136, 137, 139, 140, 148, 155, 163, 165, 171, 173, 174, 176, 181
lead, 2, 8, 18, 24, 41, 59, 64, 65, 75, 76, 115, 121, 127, 128, 129, 135, 136, 146, 148, 153, 156, 157, 168, 169
lens, 84
leukemia, 93, 142, 160
Leukocytes, 127
leukotriene, 183
Leukotriene B, 133
Leukotrienes, v, 44
LH, 7, 144
life expectancy, 155
life style, 174
limbic system, 154
linoleic acid, 33, 136, 148, 175
lipid hydroperoxide, 61, 65
Lipids, v, vi, 47, 141, 162
lipofuscin, 102, 121, 123, 124, 126, 136, 152, 153
lipoic acid, 112
lipoproteins, 102, 104, 111, 136, 139, 140, 141, 176, 179
Lipoxygenase, 33
liver failure, 68, 100, 112, 149, 161
LK, 44
Low density lipoprotein, 51, 98, 103, 105, 111, 136, 137, 139, 140, 148, 155, 163, 165, 171, 173, 174, 176, 181

lung diseases, 144
lupus erythematosus, 92
Lymphocytes, 82, 128
Lysozyme, 44

M

macrophages, 27, 40, 42, 43, 46, 50, 51, 105, 130, 136, 137, 145, 148, 163, 180
Macrophages, 127
magnetic dipole, 9
magnetic field, 9, 110
malaria, 55
malnourishment, 102
malondialdehyde, 34, 47, 48, 85, 137, 146, 175
manganese, 27, 65, 91, 166
Manganese, 116
manufacturing, 7, 31, 47
MAO, 152
marketplace, 158
matrix, 43, 131
Maxwell, 174
MC, 176
MDA, 34, 47, 48, 68, 85
mean, 80, 123, 124, 141
measurement, 9, 48, 49, 94, 124, 132, 139, 180
Melanogenesis, 77
melanoma, 76
melatonin, 33, 110, 111, 126, 151
membrane, 8, 35, 40, 41, 44, 46, 50, 51, 52, 54, 55, 59, 60, 61, 63, 80, 87, 99, 100, 102, 108, 111, 112, 134, 143, 145, 146, 162, 177, 182
mental ability, 166
MEOS, 161
metabolism, 5, 7, 10, 16, 18, 19, 39, 47, 50, 55, 56, 61, 97, 99, 101, 104, 105, 110, 113, 123, 124, 125, 127, 131, 143, 150, 152, 153, 154, 155, 161, 173, 175, 177, 182
metabolites, 25, 33, 55, 57, 79, 80, 127, 128, 135, 150, 177
metallic ions, 8, 9, 24, 30, 31, 34, 63, 64, 90, 92, 104, 151
metallothionein, 113, 115, 181, 183
methemoglobinemia, 54
methodology, 9, 178
methyl, 2, 97, 101, 143
MI, 139
Middle East, 170
mining, 144
mitochondria, 12, 25, 27, 29, 34, 35, 36, 37, 38, 39, 54, 91, 93, 95, 102, 174
mitochondrial, 12, 13, 14, 16, 35, 38, 51, 91, 97, 111, 123, 125, 127, 150, 153, 180

mitochondrial respiration, 14, 16, 24, 35, 37, 38, 51, 97, 120, 125, 127
mitosis, 52, 83, 84, 108
mixed disulfides, 114
MLSP, 123, 125
Mn, 23, 24, 91, 153, 167
mode, 158
model, 142, 181
models, 54, 98, 130, 160
molecular, 14, 22, 24, 30, 53, 58, 66, 73, 79, 80, 81, 95, 96, 112, 138, 151, 177, 182
molecular weight, 14, 58, 60, 66, 79, 95, 96, 112, 151
monocytes, 27, 41, 128
multiple sclerosis, 64
muscles, 134
muscular dystrophy, 92
myocardial infarction, 109, 136
Myoglobin, 137

N

Nacetylcysteine, 159
NADPH oxidase, 41, 147
natural, 16, 18, 33, 47, 50, 51, 72, 74, 80, 81, 87, 88, 90, 107, 110, 113, 114, 118, 125, 128, 133, 143, 165, 167, 170, 171, 180
necrosis, 52, 60, 73, 76, 83
nervous system, 15, 138
neurotransmitter, 14, 21, 151
neutral, 2, 14
neutrophiles, 27, 41, 178
nicotinamide adenine dinucleotide, 4, 12, 14, 15, 37, 38, 66
nicotinamide dinucleotide phosphate, 37, 38, 42, 94, 98
nitrates, 54, 75
nitric oxide, 9, 15, 21, 113, 137, 138, 148, 153, 183
Nitric oxide, 9, 21, 22, 138, 180
nitroderivatives, 18, 19
nitrogen, 7, 9, 21, 73, 75, 76, 95, 144
NMDA, 153
NO, 5, 9, 21, 22, 52, 74, 75, 137, 138
nodes, 115
noradrenaline, 152
norepinephrine, 110
North Atlantic Council, 159
North Carolina, 14, 21
NSAIDs, 134, 156, 160, 171
nucleic acids, 7, 18, 19, 20, 21, 30, 31, 32, 53, 57, 66, 71, 73, 76, 77, 79, 81, 127
Nucleic acids, 57

O

ocular sclerosis, 85
Ohio, 115
olfaction, 138
oncogene, 180
opiates, 168
organelles, 35, 37, 38, 39
organic, 2, 4, 5, 6, 8, 20, 23, 25, 28, 30, 47, 50, 55, 57, 58, 64, 75, 81, 88, 90, 94, 95, 96, 98, 102, 108, 116, 160, 166, 168
organic compounds, 5, 23, 25, 96
organization, 163
osteoporosis, 51, 116
overload, 65, 160
Oxidases, 14
oxidative damage, 53, 123, 174, 178, 180
oxidative phosphorylation, 36, 38, 67
oxidized, 6, 11, 13, 16, 34, 35, 51, 54, 57, 61, 73, 79, 85, 94, 101, 104, 111, 118, 124, 136, 162, 163
oxidoreductive reactions, 24
oxido-reductive reactions, 1, 2, 3
Oxygen, v, vi, 15, 23, 24, 25, 51, 56, 83, 119, 146, 173, 175, 176, 177, 179, 181, 182
oxygen activation, 31, 63, 64, 83, 92, 112, 126, 129, 142, 160
Oxygen activation, 83
oxygen consumption, 19, 25, 34, 36, 37, 123, 125, 127
Ozone, 73, 74, 175
ozone hole, 74
ozone layer, 74

P

Pacific, 169
PAN, 75
pancreas, 104, 109, 154
pancreatitis, 181
Paracetamol, 101, 179
paramagnetic, 9, 26
Paraquat, 92, 159
pathological conditions, 14, 50, 129, 134, 135, 147, 153, 177
pectin, 170
Pennsylvania, 35
peptide, 113
Peptides, 41
performance, 48, 110, 114
Peroxidase, 6, 94
Peroxidation, v, 31, 47, 123, 136, 179, 183
Peroxide, v

peroxisomes, 38, 39, 40, 93, 95, 141, 175, 179
Peroxisomes, v, 38, 39, 40, 50
PES, 122
PGE$_2$, 134, 146
pH, 27, 42, 44, 64, 91, 104
phagocytosis, 40, 42, 43, 44, 46, 51, 104, 117, 131, 133, 147, 175
Phagocytosis, v, 40, 50, 145
phagocytotic leukocytes, 17, 40, 41
pharmaceuticals, 116
phenol, 18
phenothiazines, 162
phenothiozines, 160
phenylalanine, 73
phenylbutazone, 47, 134
phospholipid, 58, 61, 139, 144, 167
photochemical reactions, 24, 73, 74, 75
photon, 71
photons, 71, 80
photosensitization, 79
photosynthesis, 16, 71, 108
phototoxic effects, 73, 76
phototoxicity, 71
photsynthesis, 24
physiology, 3
phytic acid, 88, 113
phytoestrogens, 51, 113
pituitary gland, 104
PLA, 44
placebo, 103
plasma, 16, 25, 30, 34, 47, 48, 49, 59, 64, 65, 66, 67, 92, 97, 99, 100, 101, 103, 104, 106, 111, 112, 116, 120, 122, 132, 135, 136, 138, 139, 140, 141, 147, 148, 149, 150, 155, 167, 170, 171, 173, 175, 176, 179, 180, 182
plasma proteins, 112, 180
Platelets, 46, 134
PMNL, 40, 43, 51, 93, 104, 112, 117
polarization, 14
pollution, 5, 51, 73, 75, 127, 128, 179
polycyclic hydrocarbons, 18, 47
Polyphenols, 16, 118
polyphosphate, 144
polyunsaturated, 25, 31, 32, 49, 64, 66, 75, 98, 107, 109, 111, 125, 126, 131, 136, 139, 147, 151, 155, 162, 163, 164
polyunsaturated fatty acids, 25, 31, 32, 49, 64, 66, 75, 98, 111, 125, 126, 131, 136, 147, 151, 155, 162, 163
polyunsaturated fatty acids (PUFA), 25, 32, 34, 49, 50, 164, 169
population, 74, 106, 121, 163, 164
porphyrin, 108

Potassium, 170
Prague, 121
principle, 6, 12, 45, 53, 73, 79, 95, 97, 99, 102, 150, 151
production, 1, 19, 26, 28, 29, 30, 31, 34, 36, 37, 38, 39, 41, 42, 46, 50, 54, 56, 61, 65, 66, 67, 72, 73, 75, 77, 79, 87, 91, 93, 95, 98, 105, 110, 113, 114, 119, 122, 127, 137, 143, 144, 145, 146, 147, 150, 152, 155, 161, 166, 181
prolactin, 134
proliferation, 16, 108, 114, 136, 142, 143
prooxidant, 64, 65, 88, 98, 104, 112, 121, 124, 161, 178
prooxidative drugs, 158
Propolis, 167
prostaglandin, 44, 178
Prostaglandin, 47, 50, 133
Prostaglandins, v, 44, 132, 156
protective, 7, 30, 35, 39, 48, 59, 74, 77, 80, 81, 99, 103, 105, 109, 117, 119, 144, 151, 168, 181
protein, 20, 24, 30, 37, 38, 43, 50, 59, 61, 65, 73, 79, 85, 97, 111, 112, 132, 135, 139, 141, 148, 155, 167, 168, 174, 182
proteins, 5, 7, 18, 20, 21, 22, 25, 29, 30, 32, 35, 42, 50, 53, 58, 64, 65, 71, 73, 76, 79, 81, 82, 90, 92, 97, 98, 108, 112, 113, 117, 120, 128, 132, 135, 137, 138, 154, 155, 167, 168, 177
proton, 4
psoriasis, 26, 71, 79
public health, 135
Pulmonary edema, 147
purine, 7, 19, 81, 111, 154

Q

Quinones, v, 16, 28, 33, 97, 158

R

Radiation, vi, 71, 80, 81, 183
radiation exposure, 82
radiation sickness, 20, 80, 81, 82, 83
radioactivity, 80
radiolysis, 4, 26, 31, 81
radiosensitizers, 19, 84, 105
rancidification, 3, 7
rapid growth, 101
RDA, 103, 106, 109, 115, 116
reaction time, 32
reactive oxygen species, 17, 24, 25, 159
reactive oxygen species (ROS), 14, 17, 20, 24, 25, 159

reasoning, 111
receptor, 21, 46, 51, 109, 116, 137, 151, 153
receptors, 40, 109, 136, 146, 168
recognition, 40
redox potential, 11, 19, 36
redox regulation, 182
reduced form of NADP, 6, 15, 37, 41, 42, 54, 55, 56, 57, 67, 94, 98, 147
repair, 80, 81, 82, 109, 129, 142, 154, 178, 181
reproduction, 116
resistance, 55, 56, 103, 114, 115, 125, 127, 128, 138, 147, 148, 150, 153, 154, 157, 182
respiratory diseases, 75, 145
respiratory tract, 144
response, 1, 40, 42, 43, 65, 103, 121, 128, 143, 145, 148, 174
retina, 25, 85, 110, 155
retinoic acid, 51, 109
Retinoids, 109
retinopathy, 154
retrieval, iv
retrolenticular fibroplasia, 85
rheumatic conditions, 132
rheumatism, 132, 135, 163
rheumatoid arthritis, 115, 132, 165
riboflavin, 13, 113
risk factor, 98, 131, 136, 139, 140, 141, 176, 182
risk factors, 136, 140, 176, 182
Romania, 92
ROOH, 6, 31, 89, 94, 102

S

salicylate, 153
salts, 28, 135, 159, 168
sample, 43
saturated fats, 162, 163
saturation, 106, 125
SC, 175, 181
scavengers, 59, 155
Schiff base rearrangement, 155
Schiff bases, 121, 123
selenium, 92, 95, 113, 114, 115, 155, 169, 170, 171, 180
self, 52, 61, 89
semiquinones, 4, 12, 13, 14, 16, 28, 50, 51, 57, 78
senescence, 101, 124, 180
sensitivity, 29, 47, 48, 55, 77, 82, 97, 104, 120, 121, 123, 128, 154, 163
septic shock, 15, 138
series, 110
serotonin, 13, 46, 90, 130, 146, 154
serum, 65, 105, 163, 174, 180

shellfish, 116
sickle cell anemia, 55
silicosis, 144, 145, 175
singlet, 25, 26, 31, 67, 72, 73, 77, 78, 79, 80, 89, 108, 112
singlet oxygen, 25, 26, 31, 67, 72, 73, 77, 78, 79, 80, 89, 108, 112
skin cancers, 77, 79
smog, 5, 74, 75, 76
smoking, 111, 136, 139, 140, 141, 144, 145, 148, 149, 168, 178
smooth muscle, 113, 136, 137, 183
Sodium, 91
Sodium deoxycholate, 91
South Dakota, 114
Southeast Asia, 169
Spain, 140
specific metabolic rate (SMR), 123
spectroscopy, 8, 20
spin inversion, 24
SQs, 11
stability, 1, 2, 14, 16
stages, 100, 110
starch, 167
steroid, 109
steroids, 32, 115
stimulants, 43, 145
storage, 90, 118
Streptonigrin, 159
streptozotcyn, 154
stress, 50, 52, 53, 55, 56, 57, 61, 62, 63, 85, 98, 106, 109, 110, 111, 119, 120, 121, 122, 124, 125, 126, 128, 139, 147, 148, 152, 157, 167, 174, 175, 176, 178, 180, 181, 182
structural components, 1, 47
substrate, 36, 38, 41, 44, 66, 73, 91, 94, 97, 99, 102, 139, 146, 162
Suggestion Scheme, 173, 178, 180
sulfate, 91
sulphydryl groups, 55, 58, 85, 99, 100, 104, 112, 135, 138, 154, 159, 168
Superoxide, v, 25, 26, 27, 33, 44, 51, 52, 90, 92, 97, 107, 116, 134, 154, 173, 179, 181, 183
superoxide dismutase, 27, 37, 41, 42, 49, 57, 58, 60, 90, 91, 92, 93
Superoxide dismutase, 27, 33, 92, 97, 116, 134, 154, 179, 183
Superoxide Dismutase (SOD), 90
Sweden, 53
Switzerland, 166
symptoms, 82, 93, 109, 115, 130, 135, 145, 158, 161
synergism, 183

T

Tanaka Kakuei, 182
TBARS, 48, 50, 58, 59, 138, 164
tetrachloride, 2, 5, 27, 99, 115, 149, 159, 160, 174, 181
theory, 11
Therapy, vi, 170
thermodynamics, 1
thiobarbituric acid, 47, 48, 50
Thiobarbituric acid, 47, 59, 139
thiobarbituric acid reactive substances, 48
thiol, 100, 175, 182
Thiols, 28
threshold, 60, 67, 68, 160
thrombosis, 105, 148
thromboxanes, 44, 45, 95, 129, 162
thylil radical, 20
Thymidine, 20, 125
T-lymphocytes, 108, 111
TM, 174
tolerance, 116
topological defects (TD), 179
toxicity, 50, 55, 61, 64, 92, 113, 114, 115, 116, 128, 146, 160, 161, 174, 178, 181
tranquilizer, 166
transcription, 112, 128, 143
transduction, 41, 138
Transferrin, 44, 92
transformation, 51, 104, 108, 152
transport, 10, 11, 12, 13, 14, 16, 18, 27, 29, 35, 40, 55, 92, 98, 106, 111, 146, 155
triphenlymethyl, 2
Tryptophan, 73
tuberculosis, 84
tumor necrosis factor, 145, 148
TX, 44
Tyrosinase, 6
tyrosine, 21, 73, 77, 78, 152, 161

U

ubiquinone, 16, 111
Ukraine, 142
Ulcer, 156
United States, iv, 106, 121, 158, 162, 163
universal, 47, 123, 158
uracil, 124
uranium, 80
uric acid, 33, 66, 88, 90, 111, 161, 166
Uric acid, 38, 39, 44, 111, 181
urine, 20, 34, 48, 81, 120, 124, 182
UV, vi, 4, 20, 26, 27, 31, 68, 71, 73, 74, 75, 76, 77, 79, 84, 85
UV radiation, 26, 71, 73, 74, 75, 76, 79, 84

V

valence, 65
variable, 2, 6, 75, 92, 95, 102, 145, 175
vascular permeability, 59, 60, 112
vasodilation, 113
vegetable, 107, 110
vegetarian diet, 168, 169
vinyl chloride, 149
vitamin A, 88, 107, 108, 109, 141, 170, 171, 178
Vitamin A, 107, 108, 109, 169, 174
vitamin C, 61, 65, 88, 102, 103, 104, 105, 106, 107, 117, 138, 140, 141, 151, 169, 170, 171, 178, 181
Vitamin E, 56, 59, 61, 88, 89, 90, 102, 169, 175
Vitamin K, 16, 159
Vitamins, 90, 178, 181
VLDL, 140, 165
VM, 174
von Willebrand, 174

W

water radiolysis, 20, 26, 81, 83
wavelength, 4, 71, 73
Western Europe, 79, 169
WHO-MONICA project, 140
Wilson disease, 113, 116
workers, 144
World Health Organization, 170
Wurster red, 16

X

xanthene oxidase, 14, 105, 120
Xanthine oxidase, 15, 66, 181
xenobiotics, 40, 54, 61, 104, 114, 144, 149, 177, 178, 179

Z

zinc, 27, 91, 113, 115, 132, 159, 166
Zinc, 115, 169, 180
zinc deficiency, 115